WHISPERS OF WISDOM

The Elixir of Mastery

OrangeBooks Publication

Smriti Nagar, Bhilai, Chhattisgarh - 490020

Website:**www.orangebooks.in**

© **Copyright, 2023, Author**

All rights reserved. No part of this book may be reproduced, stored in a retrieval system, or transmitted, in any form by any means, electronic, mechanical, magnetic, optical, chemical, manual, photocopying, recording or otherwise, without the prior written consent of its writer.

First Edition, 2023

Whispers of Wisdom

The elixir of mastery

Vidhu Nair

OrangeBooks Publication
www.orangebooks.in

An Ode of Gratitude

Like a quill dipped in grace, I inscribe my sincerest gratitude upon these pages, forever indebted to some wonderful people for their love. In the gentle embrace of these pages, I find solace as I pour forth heartfelt gratitude to those who have shaped my path and kindled the flames of inspiration within my soul. Let my words dance softly like a spring breeze, carrying whispers of appreciation to each cherished name.

Dr. Pavan Sudhir, Head of the Department of Art and Aesthetics, your guidance has been a beacon of light, leading me through the corridors of new ideas and illuminating the path of art-integrated learning. Your wisdom has unfurled my wings, allowing me to soar to new creative heights.

To my parents, whose hearts beat with the rhythm of education, I offer profound thanks. Though my teenage years found me immersed in the realm of engineering, you nurtured my passion for teaching, granting me the freedom to follow my calling. In your love, I found the courage to embark on this extraordinary journey.

My dear wife, Santhi, your unwavering support and tender care have been the lighthouse in the dark nights I spent weaving the words of this book. With every gentle touch, you infused my weary spirit with renewed strength, granting me the solitude I needed to bring these thoughts to life. In your love, I discovered the true meaning of harmony.

To my beloved son, Gopi Krishnan, your innocent eyes and boundless curiosity awakened the dormant writer within me. Your unwavering belief in your lazy father's potential breathed life into the pages of this book. Your presence reminds me that our dreams are worth pursuing, no matter the obstacles we face.

Dr. P. Pramod (Director SCOLE Kerala), Dr. Anil Kumar S. (Former Research Officer, SCERT Kerala, and Sri. B. Aburaj (Director, SIEMAT Kerala), you believed in my abilities and entrusted me with the freedom to blossom as a teacher and leader. Your faith in my vision allowed me to explore uncharted territories, shaping me into a better version of myself. I am forever grateful for your unwavering support.

To my dear friends, Ms. Naimisha Parmar, Mr. Maheshkumar M, Ms. Nisha Santhosh, Mr. Jayakumar S, Mr. NC Vijayakumar, Ms. N Maya, Mr. Sajju\ C. Mathew, and KV Baiju, you whispered words of encouragement, igniting the fire of courage within me. Your presence in my life has been a constant source of inspiration and motivation.

Mr. Manu AC, AEO Kalloorkad, you stood as a guiding light and patronage during my tenure as Headmaster of NSS LP School Kappu. Your unwavering support empowered me to embark on innovative endeavors, allowing the seeds of change to blossom within our school's walls.

To Mr. PG Madhusoodanan Nair and all the esteemed members of the management committee of NSS LP School Kappu, your trust and faith in my vision opened doors to boundless possibilities. Your support and freedom allowed me to sow the seeds of innovation, making our school a nurturing haven for growth.

To my esteemed colleagues, Sri. Sunilkumar Thattakuzha, Ms. AS Pravitha, Ms. K. Geetha, Mr. MA Sadique, Ms. V. Bijaly, Ms. Remya V. Nair, Ms. Geethu Ganesh, Ms. P. Greeshma, Ms. Athira C. Uthaman, Mr. K. Kiran das, Ms. Anjaly Raj, Ms. Sruthi Saji, Ms. Jismal Muhammed, Mr. Francis Mathew, Ms. Chinnumol Raju, Ms. Sajitha Siva Sivan, Ms. Raji Satheesh, Ms. Biji Haridas, Ms. Renju Sajimon, Ms. Abitha Aliyar and Ms. Fithamol Basheer, you adorned my journey with collaboration and camaraderie. Together, we dared to explore the realms of imagination and brought forth the wonders of our crazy ideas into the realms of education.

To Ms. Gigi CJ, your unwavering support in freeing me from clerical duties allowed my thoughts to flow freely onto these pages. Your dedication and understanding will forever be cherished.

In this symphony of gratitude, let these gentle words be a tribute to each remarkable soul who has graced my life's journey. May the echoes of appreciation resonate forever, intertwining our spirits and reminding us of the immeasurable power of love, support, and kindness.

As I pen down these sentiments, I am humbled by the abundance of blessings that have enriched my path. 'Whispers of Wisdom' have echoed through my heart, inspiring this book, a testament to the beauty of the human spirit and the remarkable connections we forge along the way

Dedicated to

my beloved kids of
NSS LP School Kappu,

Dream big, my dear ones, and let your aspirations soar higher than the highest stars.

Love,
Vidhu Nair

CONTENTS

CANTO ONE
Prelude of Possibilities

In the vast expanse of human existence, a symphony of possibilities awaits its grand overture. This chapter sets the stage for the journey ahead, unravelling the tapestry of 21st-century skills and their profound significance in our ever-evolving world. Like a prelude, it introduces the themes, melodies, and aspirations that will resonate throughout the pages to come.

Within the fabric of this prelude, the seeds of potential lie dormant, awaiting nurturing hands and fertile minds. It is here, at the dawn of our exploration, that we pause to reflect upon the rapid pace of change and the shifting landscapes of our globalized society. The digital age has bestowed upon us a wealth of information and inter connectedness, presenting both unprecedented challenges and boundless opportunities.

As we delve deeper into the essence of 21st-century skills, we uncover their transformative power. No longer confined to traditional notions of education, these skills encompass an embroidery of competencies that extend far beyond the realms of academia. Critical thinking, creativity, collaboration, communication, adaptability, and cultural competence emerge as the pillars upon which the modern world rests.

With each passing day, the world becomes increasingly interconnected, interdependent, and complex. The challenges we face demand a new set of skills, a new way of thinking, and a new perspective on learning. The prelude beckons us to embrace the spirit of curiosity, to venture into uncharted territories, and to cultivate the skills necessary to navigate this ever-changing landscape.

But what lies at the heart of these skills? How do they shape our individual growth, societal progress, and collective consciousness? The prelude invites us to explore the intricacies of these questions, guiding us towards a deeper understanding of their significance.

Within the prelude's verses, we find stories of resilience and innovation, tales of triumph against adversity, and the wisdom of ancient civilizations. It is here that epics like the Mahabharata, the Bible, the Quran, and the Bhagavad Gita whisper their timeless lessons, revealing instances of 21st-century skills in practice since time immemorial. These narratives remind us that the foundations of these skills are deeply ingrained in our shared human heritage.

As we journey through the prelude, we encounter the echoes of history's voices, resonating with the footsteps of those who came before us. The lessons of great civilizations, the struggles of past generations, and the discoveries of visionary minds all offer glimpses into the repositories of skills woven across time. From ancient Egypt to the Renaissance, from the scientific revolution to the digital age, history paints a vivid portrait of how these skills have shaped the course of humanity.

Yet, the prelude also invites us to witness the brilliance of contemporary minds, the scientists, the thinkers, and the pioneers who embody the essence of 21st-century skills. Through their endeavor's, we see the tangible impact of these skills on innovation, progress, and the betterment of society. Their stories serve as beacons of inspiration, illuminating the boundless possibilities that lie before us.

As this prelude draws to a close, its final notes resound with a call to action. It compels us to embrace the symphony of possibilities that awaits, to nurture and develop these essential skills within ourselves and within the generations to come. It beckons us to embark upon a transformative journey, where the elixir of mastery awaits, ready to unlock our true potential and shape a future that is brimming with promise.

In the chapters that follow, we will explore the nuances of integrating 21st-century skills in education, the practical pedagogies that foster their growth, and the remarkable instances from epics, history, and the lives of scientists that illuminate their significance. Together, we shall unveil the secrets and intricacies of these skills, empowering ourselves and future generations to become agents of change in a world ripe with possibilities.

The prelude has set the stage, and the symphony of possibilities is about to begin. Let us venture forth, guided by whispers of wisdom, and unleash the elixir of mastery that resides within us all.

Canto Two

Illuminating Talents

In the depths of the human spirit, there lies an untamed flame, waiting to be ignited. It is a flame that holds within it the power to illuminate the darkest corners of our existence, to kindle the embers of our potential, and to transform the ordinary into the extraordinary. This chapter delves into the essence of these talents, exploring the multifaceted dimensions of 21st-century skills and their profound significance in shaping our lives and the world around us. Prepare to be captivated by tales of untold brilliance, as we embark on a journey that will unveil the secrets of illumination.

Once upon a time, in a small village nestled amidst rolling hills, lived a young girl named Maya. Maya possessed a natural curiosity, a hunger for knowledge that burned fiercely within her. As she embarked on her educational journey, she encountered teachers who recognized her insatiable thirst for learning. They nurtured her talents, encouraging her to think critically, question the status quo, and explore the realms of her imagination. Maya's story became a beacon of hope, a testament to the transformative power of 21st-century skills in the hands of those who dared to dream.

In a bustling metropolis halfway across the world, a team of scientists gathered in a state-of-the-art laboratory. They were on the brink of a ground-breaking discovery that would revolutionize the field of medicine. Their success, however, was not merely the result of individual brilliance but a seamless collaboration that brought together diverse perspectives, expertise, and cultural backgrounds. Through open communication, shared goals, and a deep respect for one another's ideas, they unleashed the full potential of their collective intelligence. This tale of scientific brilliance showcased the power of collaboration and cultural competence in unlocking the mysteries of the universe.

On the pages of history, we find stories of resilience, ingenuity, and triumph against all odds. One such tale takes us back to the ancient civilization of Egypt, where the great pyramids stand tall, a testament to human ingenuity. The construction of these architectural marvels required meticulous planning, precise mathematical calculations, and a deep understanding of engineering principles. The ancient Egyptians possessed the skills of critical thinking, creativity, and adaptability, as they overcame numerous challenges to leave behind a legacy that continues to inspire awe and wonder to this day.

In the realm of literature, the works of Shakespeare have mesmerized audiences for centuries. Through his plays and sonnets, Shakespeare demonstrated the power of language to evoke emotions, provoke thought, and transcend the barriers of time and space. His mastery of communication allowed him to capture the human experience with unparalleled depth and beauty. Shakespeare's legacy serves as a reminder of the enduring significance of effective communication in connecting hearts and minds across generations.

As we navigate the complexities of the 21st century, the need for these illuminating talents becomes ever more apparent. In a world of rapid technological advancements, disruptive innovations, and shifting global dynamics, the acquisition and mastery of 21st-century skills become paramount. They empower individuals to adapt, innovate, and thrive in a landscape that demands constant evolution.

In the chapters that follow, we shall delve deeper into the manifestations of these skills in epics, history, the lives of scientists, and their integration within language, mathematics, and science classrooms. Through an assortment of short stories, news pieces, and thought-provoking anecdotes, we will witness the transformative power of 21st-century skills in action. Each tale will serve as a testament to the indomitable spirit of human potential, igniting the flame of inspiration within us all.

Prepare to be enthralled by stories of triumph, to be enchanted by the brilliance of minds that have shaped our world, and to be inspired to cultivate your own illuminating talents. For within the depths of your being lies a fire waiting to be kindled, ready to illuminate the path to greatness. Embrace the journey, for it is through the illumination of our talents that we shall usher in a future brimming with boundless possibilities.

Canto Three
Mythical Chronicles, Skilful Verses

Epics of Enlightenment

Within the ancient texts and tales of the world's greatest epics, lie profound lessons and timeless wisdom that transcend the boundaries of time and culture. These epics, be it the Mahabharata, the Bible, the Quran, or the Bhagavad Gita, hold within them narratives that illuminate the significance of 21st-century skills. In this chapter, we embark on a journey through captivating stories, fables, and examples from the lives of scientists, delving into the depths of their teachings to uncover the hidden gems that inspire and guide us in our quest for enlightenment.

The Mahabharata: A Tale of Ethical Dilemmas and Critical Thinking

The Mahabharata, an ancient Indian epic, is a treasure trove of wisdom that explores complex ethical dilemmas and the necessity for critical thinking. At the heart of this epic lies the tale of the great Kurukshetra war, where two factions of cousins, the Pandavas and the Kauravas, engage in a fierce battle for the throne.

The character of Arjuna, a skilled archer and warrior, serves as a prime example of grappling with ethical dilemmas and the need for critical thinking. As the battle draws near, Arjuna finds himself standing on the battlefield, surrounded by his relatives, friends, and revered teachers who have taken the side of the Kauravas. Overwhelmed by the prospect of killing those he holds dear, Arjuna is filled with doubt and moral uncertainty.

At this moment of crisis, Arjuna seeks guidance from Lord Krishna, his charioteer and divine mentor. What unfolds is a profound dialogue between Arjuna and Krishna, known as the Bhagavad Gita. It is within this discourse that the essence of critical thinking and ethical decision-making is revealed.

Lord Krishna imparts timeless wisdom to Arjuna, urging him to analyse the situation from a broader perspective. Krishna emphasizes the importance of discernment, encouraging Arjuna to contemplate the nature of duty, righteousness, and the consequences of his actions. Through their dialogue, Arjuna learns to navigate the complexities of ethical dilemmas, weighing the various factors at play and making decisions based on moral principles rather than personal bias.

The Bhagavad Gita, within the context of the Mahabharata, showcases the significance of critical thinking as a tool for ethical decision-making. It teaches us to pause, reflect, and carefully consider the consequences of our choices. Engaging in thoughtful analysis and discernment, we can steer moral quandaries with wisdom and integrity.

The Mahabharata, an ancient Indian epic, indeed contains numerous instances that highlight the themes of collaboration, teamwork, and effective communication. Here are a couple of examples from the storyline that illustrate the power of collaboration among the Pandavas:

The Chakravyuha Formation:

During the Kurukshetra War, the Kaurava army deploys the formidable Chakravyuha battle formation, which is difficult to penetrate. Arjuna, one of the Pandava brothers and a skilled archer, possesses the knowledge to enter and navigate through the formation successfully. However, he realizes that only he knows the strategy to penetrate but not how to exit it. Recognizing the need for collaboration, Arjuna instructs his wife, Subhadra, to convey this critical information to their son, Abhimanyu, before he is born.

Abhimanyu, having learned the strategy while in Subhadra's womb, becomes the only warrior capable of entering the Chakravyuha. The other Pandavas, aware of Abhimanyu's skills and the importance of breaking through the formation, form a protective shield around him as he charges into the Chakravyuha. This instance demonstrates collaboration and effective communication, as the Pandavas collectively utilize their strengths to support Abhimanyu, leveraging their combined abilities to achieve a common goal.

The Incognito Year in Exile:

As part of their exile, the Pandavas are required to spend a year incognito. During this time, they must hide their true identities and live unrecognized among the general population. Each brother takes on a different role and profession to avoid suspicion and maintain their anonymity.

To ensure their collective safety and prevent their true identities from being discovered, the Pandavas establish a set of rules and codes of conduct among themselves. They agree to meet at a specific location at the end of the year, allowing them to reconnect and consolidate their experiences.

This instance showcases the power of collaboration and teamwork, as the Pandavas effectively coordinate and communicate with each other to navigate the challenges of their incognito year. By adhering to their agreed-upon plan, they maintain unity, support one another, and successfully reunite at the end of the designated period.

These instances from the Mahabharata highlight how the Pandavas exemplify collaboration, teamwork, and effective communication. By pooling their resources, knowledge, and skills, the Pandavas overcome obstacles and work together towards their common goals, showcasing the power of unity and synergy in achieving success. Furthermore, the Mahabharata presents numerous other instances that highlight the relevance of 21st-century skills. Characters such as Bhishma exemplify traits like leadership, strategic thinking, and adaptability. Bhishma, a respected and wise elder, demonstrates the ability to deal with complex political situations, make difficult choices, and adapt to changing circumstances.

The Mahabharata provides rich narratives and profound teachings that shed light on the significance of 21st-century skills. It imparts lessons in critical thinking, ethical decision-making, leadership, collaboration, and adaptability. Through the timeless wisdom of this epic, we are reminded of the importance of these skills in our personal and professional lives, guiding us towards a future filled with moral integrity, resilience, and growth.

The Bible: Lessons in Collaboration and Resilience

The Bible, a sacred text for millions of people around the world, offers a wealth of stories that highlight the power of collaboration and resilience. These narratives provide valuable insights into the essence of 21st-century skills and their relevance in navigating the complexities of life.

The story of Noah and the Ark, found in the book of Genesis in the Bible, showcases Noah's obedience to God's command and his commitment to the task of building an ark. According to the narrative, humanity had become corrupt and wicked, causing God to decide to send a great flood to cleanse the earth. God chose Noah, a righteous man, to build an ark and gather a diverse group of animals to survive the impending deluge. God provided Noah with specific instructions on the dimensions and design of the ark, as well as the need to bring pairs of every kind of animal on board. Noah obediently followed these instructions, demonstrating his willingness to collaborate with a higher power.

Noah, with the help of his sons, diligently worked on constructing the massive ark. Despite facing scepticism and ridicule from others who did not believe in the impending flood, Noah remained steadfast in his commitment to the task at hand. As Noah completed the ark, animals began to arrive, guided by divine providence. Noah welcomed them on board, organizing them according to their kinds. This act of gathering and caring for the animals demonstrates Noah's dedication to fulfilling God's command and preserving the diversity of life. When the ark was ready and all the animals were aboard, the floodwaters began to engulf the earth. Noah and his family, along with the animals, found refuge inside the ark, protected from the destructive forces outside. Throughout the story, Noah's obedience to God's command and his unwavering commitment to building the ark and gathering the animals exemplify his willingness to collaborate with a higher power and fulfil a greater purpose. Despite facing challenges and scepticism, Noah trusted in God's guidance and played an essential role in the survival of both humanity and the animal kingdom.

Building the ark was no small feat. It required careful planning, organization, and the involvement of Noah's family members. Together, they collaborated, pooling their skills and resources to construct the massive vessel that would provide safety and preservation for all living beings. This story emphasizes the power of collective efforts, as Noah's family worked hand in hand to overcome the challenges they faced. The tale of Noah and the Ark underscores the importance of collaboration, as Noah actively cooperated with God's plan and carried out his assigned responsibilities, thereby demonstrating the power of aligning oneself with a greater purpose.

David, a young shepherd boy, found himself facing a formidable challenge when the Philistine army, led by their champion warrior Goliath, challenged the Israelites to send out a champion to fight him. Goliath was a giant, well-armoured and skilled in battle, while David was a mere shepherd. Despite the overwhelming odds, David stepped forward, driven by his resilience and courage. His determination and belief in his abilities stemmed from his trust in God. David had honed his skills as a shepherd, protecting his flock from predators, and had developed proficiency with a slingshot.

When David faced Goliath on the battlefield, the giant warrior mocked him. But David remained resilient and steadfast. Drawing upon his skills and his trust in God, he carefully selected a stone and loaded it into his slingshot. With precise aim and unwavering resolve, David released the stone, which struck Goliath on the forehead, causing him to fall to the ground. David's victory over Goliath symbolized the triumph of resilience and determination over seemingly insurmountable odds.

Despite being faced with an imposing opponent, David remained resilient. He refused to be deterred by the size, strength, or reputation of his adversary. Instead, he drew upon his inner strength and persevered, displaying resilience in the face of adversity. David's willingness to confront Goliath, despite his youth and lack of experience as a warrior, showcased his courage. He embraced the challenge and stepped forward to defend his people, demonstrating bravery and a willingness to take risks. David employed critical thinking skills by assessing the situation and identifying a unique strategy. Recognizing that he couldn't defeat Goliath in hand-to-hand combat, he utilized his knowledge of his own skills and the terrain to his advantage, opting for a ranged attack with his slingshot. David's use of a slingshot as a weapon showcased his problem-solving abilities. He utilized the resources available to him and employed innovative thinking to overcome the challenge before him. David's faith and trust in God were integral to his resilience. He believed that God would provide him with the strength and guidance needed to overcome Goliath. This trust served as a source of inspiration and resilience throughout the encounter.

The story of David and Goliath serves as a timeless example of the importance of resilience and other 21st-century skills. David's unwavering determination, courage, critical thinking, problem-solving abilities, and faith in God enabled him to triumph against all odds, showcasing the power of these skills in overcoming challenges and achieving success. The Bible is replete with stories of individuals who demonstrate leadership, adaptability, and empathy. The Israelites were enslaved in Egypt, enduring harsh conditions under the rule of Pharaoh. In response to their suffering, God chose Moses to lead the Israelites out of Egypt and into the Promised Land. Moses, initially hesitant to take on the task, eventually embraced his role as a leader and began his mission. He confronted Pharaoh, delivering God's message to let the Israelites go. However, Pharaoh repeatedly refused to release them, resulting in a series of devastating plagues sent by God.

As the plagues escalated, Moses continued to communicate with Pharaoh, relaying God's instructions and warnings. His effective communication skills allowed him to convey messages of consequence and negotiate on behalf of the Israelites. When Pharaoh finally relented after the tenth plague—the death of the firstborn—Moses and the Israelites seized the opportunity to leave Egypt. As they journeyed through the wilderness towards the Promised Land, Moses faced numerous challenges that required him to adapt his leadership approach. When the Israelites encountered the impassable Red Sea with the pursuing Egyptian army closing in, Moses displayed his visionary leadership. He communicated with God, who instructed him to stretch out his staff over the sea. Moses followed this guidance, and miraculously, the waters of the Red Sea parted, creating a path for the Israelites to cross safely. As they continued their journey, Moses faced other obstacles, including scarcity of food and water. In each instance, he communicated with God, who provided guidance and solutions. Through Moses' effective communication with both God and the Israelites, they overcame these challenges, highlighting his leadership and adaptability in the face of adversity.

Moses also strove to inspire a sense of unity among the Israelites. He mediated disputes, settled conflicts, and established a system of governance based on God's commandments. Moses' leadership fostered a collective identity and a shared purpose among the Israelites, forging unity and resilience as they journeyed towards the Promised Land.

The story of the Exodus from Egypt exemplifies Moses' visionary leadership, effective communication, adaptability, and ability to inspire unity. He navigated challenging circumstances, communicated effectively with both God and the Israelites, adapted his leadership approach, and fostered a sense of shared purpose and unity among the Israelites throughout their journey.

Jesus Christ, the central figure in the New Testament, serves as an embodiment of compassion, empathy, and cultural competence. His teachings emphasize love, forgiveness, and understanding, urging his followers to treat others with kindness and respect regardless of their backgrounds or beliefs. Jesus' interactions with people from various walks of life exemplify the importance of cultural competence and the power of effective communication in building bridges and fostering harmony.

The stories within the Bible present timeless lessons and insights into the significance of 21st-century skills. They remind us of the power of collaboration, resilience, leadership, adaptability, and cultural competence in navigating the challenges and complexities of our modern world. By drawing inspiration from these biblical narratives, we can cultivate these skills within ourselves, fostering personal growth, fostering harmonious relationships, and creating a more compassionate and inclusive society.

The Quran: Communication and Cultural Competence

The Quran, the holy scripture of Islam, provides invaluable insights into the power of effective communication and cultural competence. It offers a unique perspective on the significance of 21st-century skills and their application in fostering harmonious relationships and understanding in a diverse world.

The stories from the life of the Prophet Muhammad highlight his resilience, determination, and perseverance. Despite facing numerous challenges and opposition, he remained steadfast in his mission, displaying unwavering faith, effective leadership, and a commitment to peace. The Prophet Muhammad's example continues to inspire people around the world, emphasizing the importance of resilience and perseverance in the face of adversity.

The Early Years in Mecca

When the Prophet Muhammad received his first revelations from God, he faced immense opposition and persecution from the powerful elite of Mecca. Despite the hostility and threats to his safety, he remained steadfast in spreading the message of Islam. The Prophet's resilience allowed him to endure years of persecution, demonstrating his unwavering commitment to his mission.

The Hijra (Migration) to Medina

Due to escalating persecution in Mecca, the Prophet Muhammad and his followers were forced to migrate to the city of Medina. This journey, known as the Hijra, was filled with dangers and challenges. The Quraysh tribe, seeking to prevent their departure, pursued them with the intent of capturing or eliminating the Prophet Muhammad. However, through careful planning and resilience, the Prophet successfully evaded capture, ultimately arriving safely in Medina. This event marked a turning point and laid the foundation for the establishment of a thriving Muslim community.

The Battle of Badr

One of the most significant battles during the early years of Islam was the Battle of Badr. The Prophet Muhammad and a small group of believers faced a well-equipped army from Mecca, outnumbered and seemingly outmatched. However, through his unwavering faith and leadership, the Prophet inspired his followers with courage and determination. Despite the odds, the Muslims emerged victorious, illustrating the power of resilience, trust in God, and effective leadership.

The Treaty of Hudaybiyyah

The Treaty of Hudaybiyyah, mentioned earlier, highlights the Prophet Muhammad's resilience in the face of setbacks and challenges. Initially, the negotiations faced obstacles, and the terms of the treaty seemed unfavourable to the Muslims. However, the Prophet remained patient and focused on the long-term goals of peace and harmony. Eventually, the treaty became a turning point, as it paved the way for future alliances and peaceful resolutions.

The Conquest of Mecca

After years of persecution and exile, the Prophet Muhammad returned to Mecca with an army of Muslims. Instead of seeking revenge, he displayed remarkable forgiveness and compassion. By embracing his adversaries and forgiving those who had wronged him, the Prophet demonstrated his unwavering commitment to peace and reconciliation. The Prophet Muhammad's teachings and actions emphasize the importance of open dialogue, empathy, and understanding in establishing strong connections and promoting unity among people.

The Quran encourages believers to communicate with wisdom, kindness, and patience, regardless of differences in language, culture, or belief systems. It emphasizes the value of active listening, respectful dialogue, and finding common ground as pathways to building bridges between individuals and communities.

One example that resonates strongly with the theme of cultural competence is the story of the Prophet Muhammad's journey, known as the Night Journey (Isra and Mi'raj). In this extraordinary event, the Prophet Muhammad is said to have travelled from Mecca to Jerusalem and ascended to the heavens, where he met with various prophets and received divine revelations.

This profound journey serves as a metaphorical representation of the Prophet Muhammad's ability to transcend cultural boundaries, connect with different communities, and gather wisdom from diverse sources. It symbolizes his role as a messenger who carried a universal message of peace, justice, and compassion to all people.

The Quran emphasizes the importance of treating others with respect, kindness, and empathy. It encourages believers to value diversity and to engage in meaningful dialogue to foster understanding and harmony. The Quran's teachings on communication and cultural competence lay the foundation for building bridges between individuals of different backgrounds and nurturing inclusive communities.

The Prophet Muhammad's own conduct provides numerous instances that exemplify effective communication and cultural competence. His interactions with people from different tribes and faiths demonstrate his ability to connect with individuals on a personal level, listen attentively to their concerns, and respond with compassion and wisdom. Through his example, the Prophet Muhammad sets a benchmark for believers to emulate in their own lives, striving to cultivate effective communication skills and cultural competence.

By drawing upon the teachings of the Quran and the example of the Prophet Muhammad, we can learn valuable lessons about the power of communication and cultural competence in promoting understanding, harmony, and peaceful coexistence. These skills enable us to bridge divides, build meaningful connections, and create a world where diverse cultures and perspectives are celebrated and valued.

CANTO FOUR

Time's Tapestry: Skilful Narratives

Literature, with its powerful storytelling and rich symbolism, has long been a source of profound insights and reflections on the human condition. The works of literary legends such as William Shakespeare, Leo Tolstoy, John Milton, Christopher Marlowe, George Bernard Shaw, John Keats, and others provide us with glimpses into the intricacies of human nature and the relevance of 21st-century skills.

Fyodor Dostoevsky's "Crime and Punishment": The novel follows the story of Raskolnikov, a troubled former student who commits a murder and is tormented by his guilt. Through Raskolnikov's internal struggle and subsequent redemption, readers witness the power of critical thinking and ethical decision-making. Raskolnikov's character represents the consequences of his actions and his eventual realization of the importance of empathy, remorse, and taking responsibility for one's actions. The novel highlights the significance of these skills in personal growth, moral development, and understanding the complexities of human nature.

Jane Austen's "Pride and Prejudice": The novel revolves around the Bennet family and their five daughters, particularly Elizabeth Bennet, a strong-willed and independent woman. Through Elizabeth's journey of self-discovery and her interactions with Mr. Darcy, readers witness the significance of critical thinking, open-mindedness, and the ability to challenge societal norms and prejudices. Elizabeth's character development showcases the importance of these skills in forming genuine connections, understanding others' perspectives, and breaking down barriers created by prejudice and social class.

Gabriel Garcia Marquez's "One Hundred Years of Solitude": The novel tells the multi-generational story of the Buendía family and their experiences in the fictional town of Macondo. Through magical realism and vivid storytelling, Marquez emphasizes the importance of imagination, creativity, and the ability to see beyond the ordinary. The novel encourages readers to embrace their own creativity, think outside the box, and appreciate the beauty and wonder that can be found in everyday life. It showcases the relevance of these skills in fostering innovation, adaptability, and a broader perspective in navigating the complexities of the modern world.

Chinua Achebe's "Things Fall Apart": The novel explores the life of Okonkwo, a respected warrior in an Igbo village, and the impact of colonialism on his community. Through Okonkwo's character, readers witness the importance of cultural competence, adaptability, and the ability to navigate cultural diversity. As the clash between traditional Igbo values and the arrival of European influence unfolds, the story emphasizes the significance of embracing and understanding different cultures, promoting empathy, and building bridges between communities—a reflection of the 21st-century skills necessary for global citizenship.

Virginia Woolf's "To the Lighthouse": The novel revolves around the Ramsay family and their visits to the Isle of Skye. Through the characters' inner monologues and Woolf's stream-of-consciousness narrative style, readers are prompted to develop skills such as empathy, active listening, and the ability to understand multiple perspectives. The story emphasizes the significance of effective communication, connection, and the appreciation of diverse experiences and viewpoints in building meaningful relationships. Woolf's exploration of human perception and subjectivity highlights the relevance of these skills in navigating the complexities of human relationships in the 21st century.

William Shakespeare's "Hamlet": The play delves into the tragic story of Prince Hamlet, who seeks revenge for his father's murder. Hamlet's character embodies qualities of critical thinking, introspection, and moral complexity. The play prompts readers to question appearances, ponder existential questions, and navigate moral dilemmas. Hamlet's journey emphasizes the importance of self-awareness, ethical decision-making, and the exploration of one's own identity in the modern world. Shakespeare's timeless work continues to remind us of the significance of these skills in understanding ourselves and our place in society.

Shakespeare's "Macbeth" explores themes of ambition, power, and the consequences of unchecked desires. The character of Lady Macbeth exemplifies the dangers of manipulating others and the inner turmoil that arises from guilt. Her character demonstrates the importance of ethical decision-making, self-reflection, and the recognition of moral boundaries. Romeo and Juliet, The iconic tragedy portrays the ill-fated love story of Romeo and Juliet, two young individuals caught in the midst of a feud between their families. Their passionate love and untimely deaths showcase the power of emotions, the consequences of impulsive actions, and the importance of communication, empathy, and conflict resolution. Sonnet 18, the famous sonnet, often referred to as "Shall I compare thee to a summer's day?", praises the beauty and eternal nature of love. It exemplifies the significance of creative expression, the ability to capture profound emotions in words, and the enduring impact of art on human experiences.

Leo Tolstoy's "War and Peace": The epic novel explores the lives of various characters against the backdrop of war and societal change. Through characters like Pierre Bezukhov, Prince Andrei Bolkonsky, and Natasha Rostova, readers witness the significance of resilience, adaptability, and the ability to navigate uncertainty and conflict. The characters' experiences reflect the challenges of their time and the complexities of human relationships. The story exemplifies the importance of emotional intelligence, empathy, and the ability to build meaningful connections in times of chaos. It showcases the 21st-century skills required to thrive in a rapidly changing world.

Anna Karenina, written by Leo Tolstoy in the 19th century, primarily explores themes of love, family, society, and the human condition. While the novel does not explicitly address 21st-century skills, we can identify several instances within the story that reflect the essence of these skills.

Throughout the novel, characters engage in deep introspection, reflecting on their actions, beliefs, and societal expectations. For instance, both Anna and Konstantin Levin question the nature of love, morality, and the purpose of their lives. Their introspective journeys demonstrate critical thinking skills by challenging existing norms and seeking personal truths.

The interactions between characters in Anna Karenina involve complex dialogues and exchanges of ideas. Tolstoy explores the dynamics of relationships, illustrating the importance of effective communication and collaboration for navigating personal and societal challenges. The discussions between Levin and his friends about agrarian reform, for example, highlight the need for constructive dialogue and cooperative problem-solving.

Tolstoy delves deeply into the emotional lives of his characters, emphasizing the significance of empathy and understanding. Anna's struggle with her desires and obligations, as well as her complex relationships with Vronsky and her son, requires her to navigate her emotions and make difficult choices. The novel highlights the importance of emotional intelligence in dealing with complex interpersonal dynamics.

Several characters in Anna Karenina face adversity and must adapt to changing circumstances. Levin, for instance, experiences setbacks and challenges in his personal and professional life but demonstrates resilience by adjusting his perspectives and embracing change. His ability to adapt and persevere reflects the 21st-century skills of adaptability and resilience.

Tolstoy presents ethical dilemmas throughout the novel, requiring characters to make choices that have significant consequences. For example, Anna must decide between societal expectations and her own desires, while Levin grapples with moral questions surrounding his interactions with the peasants on his estate. These ethical dilemmas highlight the importance of ethical decision-making and the complexities involved.

While Anna Karenina was written long before the 21st century, its exploration of complex human experiences and the skills required to navigate them resonate with the core principles of 21st-century skills such as critical thinking, communication, collaboration, emotional intelligence, adaptability, resilience, and ethical decision-making.

John Keats' poetry: Keats' works, such as "Ode to a Nightingale" and "Ode on a Grecian Urn," are known for their exploration of beauty, nature, and human emotions. His poetry embodies the significance of creativity, aesthetic appreciation, and emotional intelligence. Through vivid descriptions and evocative language, Keats encourages readers to tap into their emotions, explore the beauty of the natural world, and use the power of imagination to understand and express their experiences. His poetry showcases the relevance of these skills in fostering emotional well-being, creative expression, and a deeper understanding of ourselves and the world around us.

Christopher Marlowe's "Doctor Faustus": The play follows the tragic story of Dr. Faustus, a scholar who makes a pact with the devil in exchange for unlimited knowledge and power. Through Faustus' journey, readers witness the dangers of unchecked ambition and the pursuit of knowledge without considering the ethical implications. The play serves as a cautionary tale, highlighting the importance of critical thinking, ethical decision-making, and the ability to reflect on one's values and purpose. It emphasizes the need for responsible use of knowledge and the recognition of the potential consequences of our actions.

John Milton's "Paradise Lost": The epic poem delves into the biblical narrative of the fall of man, exploring themes of free will, temptation, and the struggle between good and evil. Through the characters of Satan, Adam, and Eve, readers witness the significance of moral reasoning, resilience, and the pursuit of redemption. Milton's work prompts readers to question their own choices, reflect on the consequences of their actions, and strive for personal growth and moral integrity. "Paradise Lost" exemplifies the 21st-century skills of critical thinking, moral reasoning, and the exploration of one's own beliefs and values.

John Milton's Paradise Lost, the poet explores the theme of redemption through the character of Jesus Christ. The poem emphasizes the significance of resilience, moral integrity, and the pursuit of spiritual enlightenment. It showcases the 21st-century skills of critical thinking, introspection, and the exploration of one's purpose and values. "Samson Agonistes" The tragedy focuses on the biblical character of Samson and his struggle against the Philistines. Samson's journey of self-discovery, inner conflict, and eventual redemption highlights the importance of resilience, self-reflection, and the pursuit of justice. The play emphasizes the significance of moral reasoning, determination, and the willingness to challenge oppressive systems.

These elaborations provide a more in-depth exploration of the events, characters, and themes in each book, demonstrating how they exemplify the significance of 21st-century skills such as critical thinking, creativity, empathy, cultural competence, and ethical decision-making. By engaging with these texts on a deeper level, readers can develop a richer understanding of these skills and their relevance in navigating the complexities of the modern world.

CANTO FIVE

Luminous Innovators, Enchanting Skills

In this chapter, we delve into the remarkable lives of scientists who embody the spirit of 21st-century skills—curiosity, critical thinking, collaboration, and creativity. Through their innovative discoveries and relentless pursuit of knowledge, these luminous innovators have reshaped our understanding of the world and inspired generations to come. Their stories serve as a testament to the power of these skills in driving scientific progress and shaping the future.

Isaac Newton: The Brilliant Mind Unveiled

The Falling Apple

One of the most famous stories associated with Isaac Newton is the falling apple that reportedly inspired his understanding of gravity. As the legend goes, Newton was sitting under an apple tree when an apple fell from the branch and hit him on the head. This event sparked his curiosity, leading him to contemplate why the apple fell straight down to the ground instead of moving in a different direction. This observation eventually led him to develop his theory of universal gravitation, a fundamental principle in physics that explains the force of attraction between objects.

Newton's Prism Experiments

Newton's experiments with optics played a significant role in advancing our understanding of light and colour. He conducted experiments using prisms to investigate the nature of white light and its constituent colours. By passing a beam of sunlight through a prism, Newton observed that the light refracted and split into a spectrum of colours, forming a rainbow-like pattern. This discovery led him to propose that white light is composed of a spectrum of colours, laying the foundation for the field of spectroscopy and our understanding of the electromagnetic spectrum.

Newton's Laws of Motion

Newton's laws of motion revolutionized the field of physics and provided a mathematical framework to explain the motion of objects. His three laws, known as Newton's Laws of Motion, describe the relationship between an object and the forces acting upon it. The first law states that an object at rest will remain at rest, and an object in motion will continue in motion unless acted upon by an external force. The second law relates the force applied to an object to its acceleration and mass, while the third law states that for every action, there is an equal and opposite reaction. These laws formed the basis for classical mechanics and continue to be foundational principles in physics.

Isaac Newton's life highlight his curious nature, keen observation skills, and logical reasoning abilities. They demonstrate how his relentless pursuit of knowledge and systematic inquiry led to radical discoveries that shaped our understanding of the physical world. Newton's experiences serve as powerful examples of critical thinking, systematic experimentation, and the pursuit of knowledge, emphasizing the significance of these skills in scientific exploration and discovery.

Marie Curie: Illuminating the World with Science

The Discovery of Polonium and Radium

Marie Curie's relentless pursuit of scientific knowledge led her to the discovery of two revolutionary elements: polonium and radium. Working alongside her husband Pierre Curie, Marie conducted meticulous experiments on various minerals, seeking new elements with unique properties. After years of painstaking work, they successfully isolated polonium in 1898, naming it after Marie's native Poland. Shortly after, they discovered radium, which proved to be even more significant. These discoveries not only revolutionized our understanding of radioactivity but also paved the way for advancements in medical treatments and radiation therapy.

The Mobile Radiography Units

During World War I, Marie Curie recognized the urgent need for mobile radiography units to aid in diagnosing injuries on the battlefield. With her expertise in radiation, she spearheaded the development of these units, known as "petite Curies." Equipped with X-ray machines and operated by trained medical personnel, these units provided crucial diagnostic support to doctors in the field. Marie's determination, leadership, and collaborative efforts showcased her commitment to applying scientific knowledge for the benefit of humanity, even in challenging circumstances.

The Isolation of Radium

Marie Curie's unwavering dedication to her research is evident in the isolation of radium, a process that required immense patience and persistence. Radium was present in minuscule quantities within the raw materials Marie worked with, making its isolation a formidable challenge. Through a series of intricate chemical processes and repeated refinement, Marie succeeded in extracting pure radium, even in the face of technical obstacles and the hazardous nature of the radioactive materials involved. Her resilience and unwavering commitment to her work exemplify the significance of perseverance and dedication in scientific endeavours.

Marie Curie's life underscores her intellectual courage, unwavering dedication, and collaborative mindset. They highlight her pioneering discoveries, her pioneering work in the field of radioactivity, and her enduring legacy in science. Marie Curie's accomplishments serve as inspiring examples of the importance of perseverance, open-mindedness, and interdisciplinary collaboration in scientific pursuits, and they continue to inspire generations of scientists to this day.

Albert Einstein: Unravelling the Mysteries of the Universe

The Thought Experiments

Albert Einstein was known for his extensive use of thought experiments to explore complex scientific concepts. One of his most famous thought experiments involved imagining what it would be like to ride alongside a beam of light. Through this exercise, Einstein questioned the long-held belief in the absolute nature of time and space. His imaginative thinking and willingness to challenge established ideas led him to develop the theory of relativity, which revolutionized our understanding of the fabric of the universe.

The Miracle Year of 1905

In 1905, often referred to as Einstein's "miracle year," he published four ground-breaking papers that transformed the field of physics. One of these papers introduced the theory of special relativity, which presented the radical idea that the laws of physics are the same for all observers in uniform motion. This breakthrough concept challenged the traditional understanding of space and time and paved the way for further developments in Einstein's work. The sheer creativity and intellectual flexibility demonstrated during this remarkable period highlight Einstein's ability to think outside the box and embrace unconventional ideas.

The Eddington Expedition

Einstein's theory of general relativity predicted that light would be affected by gravity. To confirm this prediction, British astronomer Sir Arthur Eddington organized an expedition to observe a solar eclipse in 1919. Einstein's theory proposed that the gravity of the Sun would bend the path of light passing near it. The successful observation of this phenomenon during the eclipse provided experimental evidence supporting Einstein's theory and solidified his status as a scientific icon. This event showcased Einstein's creative problem-solving skills and his willingness to subject his ideas to rigorous experimental testing.

Albert Einstein's life exemplifies his imaginative thinking, willingness to challenge conventional wisdom, and creative problem-solving abilities. They highlight his revolutionary contributions to the field of physics and the profound impact of his ideas on our understanding of the universe. Einstein's ability to imagine new possibilities, think critically, and pursue unconventional ideas serves as a powerful testament to the significance of imagination, intellectual flexibility, and the pursuit of knowledge beyond traditional boundaries.

Rosalind Franklin: Unveiling the Structure of Life

The X-ray Diffraction Studies

Rosalind Franklin's meticulous X-ray crystallography studies were instrumental in unravelling the structure of DNA. She conducted detailed experiments to obtain high-quality X-ray diffraction images of DNA fibres and crystals. One of her notable achievements was capturing the iconic Photo 51, an X-ray image of DNA that provided critical insights into its double-helix structure. Franklin's relentless pursuit of precise data and her ability to interpret complex X-ray patterns set the stage for the subsequent breakthroughs in understanding the DNA molecule.

The Race for DNA Structure

During her research, Franklin faced significant challenges and found herself embroiled in the intense competition to uncover the structure of DNA. At the time, there was a race between different research groups, including Franklin's team and the duo of James Watson and Francis Crick. Despite facing setbacks and limited support, Franklin remained focused on her work, emphasizing the importance of rigorous analysis and accurate interpretation of experimental results. Her determination and dedication to the scientific process were crucial in providing key data that eventually contributed to the discovery of the DNA structure.

The Legacy of Photo 51

Photo 51, the X-ray diffraction image captured by Rosalind Franklin, played a pivotal role in deciphering the structure of DNA. Although Franklin herself did not directly participate in the publication of the famous Watson and Crick paper announcing the double-helix structure, her work and the insights gained from Photo 51 were integral to their breakthrough. The image revealed the characteristic X-shaped pattern that hinted at the helical structure of DNA. Franklin's precise data and analytical approach served as crucial pieces of evidence that supported the final elucidation of the DNA structure.

These episodes from Rosalind Franklin's life emphasize her meticulous research, attention to detail, and perseverance in the face of challenges. They highlight her significant contributions to the discovery of the DNA structure, which laid the foundation for our understanding of genetics and molecular biology. Franklin's dedication to excellence, precision, and analytical thinking serves as a testament to the importance of these skills in scientific research. Her work continues to inspire scientists and underscores the significance of meticulous attention to detail and unwavering pursuit of knowledge.

Nikola Tesla: Harnessing the Power of Innovation

The Tesla Coil and Wireless Power Transmission

One of Nikola Tesla's most iconic inventions was the Tesla coil, a high-voltage resonant transformer that revolutionized the field of electrical engineering. Tesla's experiments with the coil demonstrated the wireless transmission of electricity over long distances, showcasing his visionary thinking and innovative approach. His dream was to create a world where energy could be transmitted wirelessly, providing power to remote areas and eliminating the need for traditional power lines. Although his vision of a global wireless power distribution system was not fully realized, his experiments laid the foundation for future advancements in wireless technology.

Alternating Current and the War of Currents

Nikola Tesla's work on alternating current (AC) electricity was instrumental in the "War of Currents" during the late 19th century. Tesla championed the use of AC, advocating for its superiority over Thomas Edison's direct current (DC) system. He demonstrated the efficiency and safety of AC, showcasing its ability to transmit electricity over long distances with less energy loss. Tesla's contributions to AC power systems not only transformed the way electricity was generated and distributed but also laid the groundwork for the modern power grid that powers our cities today.

The Wardenclyffe Tower and Wireless Communication

One of Tesla's ambitious projects was the construction of the Wardenclyffe Tower, intended to be a massive wireless communication and energy transmission facility. Tesla envisioned a world where information and energy could be transmitted wirelessly across vast distances. Although the project was ultimately abandoned due to funding challenges, the concept of wireless communication that Tesla pursued paved the way for future developments in radio technology and the eventual invention of the radio. His ideas and inventions continue to shape the wireless communication systems we rely on today.

Nikola Tesla's life highlights his imaginative mind, capacity for innovation, and dedication to advancing humanity's progress. They underscore his revolutionary contributions to the fields of electricity, magnetism, and wireless technology, demonstrating the significance of creativity, adaptability, and the ability to think beyond existing limitations. Tesla's work continues to inspire scientists, engineers, and inventors, reminding us of the transformative power of visionary thinking and the pursuit of trailblazing ideas.

Leonardo da Vinci: Mastering the Art of Science

The Vitruvian Man

Leonardo da Vinci's famous drawing, the Vitruvian Man, exemplifies his interdisciplinary approach and holistic thinking. In this artwork, da Vinci explored the relationship between human anatomy and geometry, combining his knowledge of anatomy with his understanding of proportion and symmetry. The drawing depicts a man with outstretched arms and legs, fitting perfectly within a circle and a square. Through this piece, da Vinci merged the disciplines of art, mathematics, and anatomy, showcasing his ability to integrate diverse fields of knowledge to create a harmonious and holistic representation of the human form. The Vitruvian Man symbolizes da Vinci's belief in the interconnectedness of different disciplines and his pursuit of a comprehensive understanding of the world.

The Last Supper

Leonardo da Vinci's masterpiece, The Last Supper, demonstrates his holistic problem-solving skills and attention to detail. The painting depicts the final meal of Jesus and his disciples, capturing their individual expressions and emotions. Da Vinci meticulously studied human anatomy and facial expressions to create realistic and expressive figures in the painting. He employed his understanding of perspective, light and shadow, and composition to convey a narrative and create a sense of depth and atmosphere. The Last Supper is a testament to da Vinci's interdisciplinary approach, combining his knowledge of art, human psychology, and storytelling to create a powerful and emotionally resonant artwork.

The Codex Leicester

The Codex Leicester, a collection of da Vinci's scientific writings and observations, reflects his insatiable curiosity and his pursuit of knowledge across diverse fields. In this codex, da Vinci explored various topics, including geology, astronomy, hydrodynamics, and the properties of light. He recorded his observations, made sketches, and posed questions that challenged established beliefs and theories. Da Vinci's interdisciplinary investigations and his relentless curiosity led him to make revolutionary discoveries and advance our understanding of the natural world. The Codex Leicester stands as a testament to his holistic thinking and his belief in the importance of exploring multiple disciplines to unravel the mysteries of the universe.

Da Vinci's life exemplifies the significance of interdisciplinary approach, insatiable curiosity, and holistic thinking. Da Vinci's ability to integrate knowledge from various fields, his relentless pursuit of understanding, and his capacity to approach problems from multiple angles allowed him to make profound contributions to art, science, and innovation. His approach serves as a timeless inspiration for individuals seeking to embrace diverse perspectives, think holistically, and tackle complex challenges by drawing upon knowledge and insights from different domains. It highlights the power of integrative skills, holistic problem-solving, and the pursuit of knowledge across diverse fields in shaping our understanding of the world and driving meaningful progress.

Jane Wang: A Trailblazer in Robotics and Artificial Intelligence

The Robotic Prosthetic Limb

Jane Wang's ground-breaking work in robotics and artificial intelligence led to the development of a revolutionary robotic prosthetic limb. Inspired by her passion for helping individuals with limb loss regain their mobility and independence, Wang embarked on a journey to bridge the gap between technology and human needs. Through interdisciplinary collaboration with engineers, neuroscientists, and medical professionals, she designed a robotic limb that not only mimicked the movements of a natural limb but also integrated advanced sensory feedback. This interdisciplinary approach, combining knowledge from engineering, neuroscience, and medicine, enabled Wang to create a prosthetic limb that enhanced the quality of life for individuals with limb loss, showcasing the significance of integrative skills and holistic problem-solving in the field of robotics.

Ethical Considerations in Artificial Intelligence

As advancements in artificial intelligence accelerated, Jane Wang recognized the importance of ethical considerations in the development and deployment of AI technologies. She delved into the ethical implications of AI, addressing concerns such as bias, privacy, and the potential impact on society. Wang's interdisciplinary mindset allowed her to engage with experts from fields such as philosophy, sociology, and law, fostering discussions on responsible AI development. Through her research and advocacy, she emphasized the need for an interdisciplinary approach that considers not only technical aspects but also societal, ethical, and legal dimensions, ensuring that AI technologies are developed and implemented with the well-being of individuals and society in mind.

Collaborative Innovation in Robotics

Jane Wang's collaborative mindset and ability to bring together diverse teams played a pivotal role in driving innovation in the field of robotics. She recognized the power of interdisciplinary collaboration and sought to create an environment where experts from different fields could come together and contribute their unique perspectives. Through collaborative projects, Wang facilitated the exchange of knowledge and ideas, encouraging engineers, computer scientists, psychologists, and designers to work together towards common goals. This collaborative approach fostered innovation, enabling the development of robotic systems that were not only technically advanced but also user-friendly, adaptable, and tailored to human needs. Wang's ability to bridge the gap between technology and human-centric design showcases the significance of interdisciplinary skills and the pursuit of knowledge across diverse fields in shaping the future of robotics and artificial intelligence.

Wang's ability to integrate knowledge from various disciplines, collaborate with experts from diverse backgrounds, and consider ethical implications highlights the importance of integrative skills, holistic problem-solving, and the pursuit of knowledge across diverse fields. Her work serves as an inspiration for future scientists and engineers, emphasizing the need to embrace interdisciplinary approaches, maintain a curious mindset, and address ethical considerations to create transformative technologies that benefit humanity and address complex societal challenges.

Stephen Hawking, The Real Fighter

Overcoming Adversity

Stephen Hawking's life is a testament to resilience and the power of the human spirit. Diagnosed with a rare form of motor neuron disease at a young age, Hawking faced immense physical challenges. Despite his deteriorating physical condition, his insatiable curiosity and determination to pursue his passion for physics remained unwavering. Through sheer willpower and the aid of technology, including his iconic speech-generating device, Hawking continued to push the boundaries of scientific knowledge. His ability to adapt to his circumstances, leveraging interdisciplinary skills and technology, showcases the significance of adaptability and the pursuit of knowledge across diverse fields.

Unravelling the Mysteries of the Universe

Stephen Hawking's ground-breaking work in theoretical physics, particularly in the field of cosmology, revolutionized our understanding of the universe. His research on black holes, the Big Bang theory, and the nature of time challenged existing scientific paradigms and opened up new frontiers of exploration. Hawking's interdisciplinary approach, combining principles from physics, mathematics, and philosophy, allowed him to develop ground-breaking theories that bridged gaps between different fields of study. His holistic thinking and ability to integrate knowledge from diverse disciplines exemplify the significance of integrative skills and holistic problem-solving in unravelling complex scientific phenomena.

Communicating Complex Science

Stephen Hawking's ability to communicate complex scientific concepts to the general public showcased his talent for bridging the gap between technology and human understanding. Despite the intricate nature of his research, Hawking possessed a gift for distilling complex ideas into accessible language, making them relatable to a wider audience. Through his books, lectures, and media appearances, he inspired millions around the world to develop an interest in science and to contemplate the mysteries of the universe. This unique skill of effective science communication demonstrated the significance of interdisciplinary skills and the pursuit of knowledge across diverse fields, enabling the dissemination of scientific knowledge and fostering a sense of wonder and curiosity in people of all backgrounds.

CANTO SIX

Lingual Melodies: Interweaving the Power of Words

Language learning is a dynamic process that goes beyond the acquisition of grammar and vocabulary. In the 21st century, there is a growing recognition of the need to equip learners with essential skills that will empower them to thrive in an interconnected and rapidly evolving world. This chapter explores the practical integration of 21st century skills in language classrooms, offering pedagogical strategies and real-life case studies that demonstrate the transformative impact of such integration.

Understanding 21st Century Skills in Language Learning

The future of language learning in classrooms is likely to be influenced by technological advancements, evolving teaching methodologies, and changing societal needs. Here are some potential predictions and strategies for the future of language learning:

In the future of language learning, personalized and adaptive learning will play a significant role in providing tailored instruction to meet the unique needs of individual learners. With the advancement of artificial intelligence (AI) and machine learning, language learning platforms and apps will be able to analyse learners' strengths, weaknesses, and learning preferences to create personalized learning experiences.

The concept of personalized learning recognizes that every learner has different abilities, interests, and learning styles. Traditional classroom settings often struggle to cater to the diverse needs of students, leading to gaps in knowledge and disengagement. However, with the help of AI, language learning platforms can adapt to the specific requirements of each learner, providing targeted instruction, practice, and feedback.

One of the primary benefits of personalized learning is its ability to identify learners' strengths and build upon them. AI algorithms can assess learners' language skills through diagnostic tests, adaptive quizzes, and language assessments. Based on the results, the platform can generate individualized learning paths that focus on areas that require improvement while allowing learners to progress at their own pace in areas of strength. For example, if a learner excels in vocabulary but struggles with grammar, the platform can provide additional grammar exercises and explanations to address the specific needs of that learner.

Adaptive learning, a key component of personalized learning, allows the language learning platform to dynamically adjust the difficulty and content of the activities based on learners' performance and progress. As learners interact with the platform, AI algorithms continuously analyse their responses, track their progress, and provide immediate feedback. This real-time feedback helps learners identify and correct their mistakes, reinforcing their understanding and boosting their confidence. For instance, if a learner consistently answers grammar questions correctly, the platform can increase the complexity of the grammar exercises to provide a challenge that matches the learner's proficiency level.

Personalized and adaptive learning platforms can employ various instructional strategies to cater to different learning preferences. Some learners may prefer visual aids, while others may benefit from audio-based activities. The AI algorithms can take these preferences into account and present the content in a manner that aligns with each learner's preferred learning style. For example, learners who are visual learners may receive more visual cues, such as infographics or diagrams, to reinforce language concepts.

Personalized learning platforms can incorporate data analytics to track learners' progress over time. This data can be used to generate comprehensive reports that provide insights into learners' strengths, areas for improvement, and progress. Teachers can leverage this information to provide targeted guidance and support, adapting their instruction based on individual learners' needs. This combination of technology-driven insights and teacher expertise enhances the learning experience and maximizes learner outcomes.

However, while personalized and adaptive learning offers many advantages, it is essential to strike a balance between technology and human interaction. Although AI algorithms can provide personalized instruction, human teachers play a crucial role in supporting and guiding learners. The future of language learning should involve a blended approach, where technology serves as a valuable tool that complements and enhances the teaching and learning process.

Personalized and adaptive learning holds immense potential in the future of language learning. Through AI and machine learning, language learning platforms can analyse learners' strengths, weaknesses, and preferences to provide tailored instruction, adaptive activities, and immediate feedback. By catering to individual learners' needs, personalized learning promotes engagement, accelerates progress, and boosts learner outcomes. However, it is crucial to maintain a balance between technology and human interaction, as educators continue to play a vital role in guiding and supporting learners on their language learning journey.

Integrating 21st-century skills into language classrooms is essential for preparing students to thrive in an interconnected and rapidly changing world. By incorporating these skills into language learning activities, educators can foster critical thinking, communication, collaboration, creativity, and digital literacy. Here's how 21st-century skills can be integrated into language classrooms:

Critical Thinking: Language learning provides an excellent opportunity for students to develop critical thinking skills. Teachers can design activities that require students to analyse and evaluate information, make connections, and solve problems. For example, students can engage in debates, conduct research, or critically analyse literary texts. By encouraging students to question, evaluate evidence, and think critically, language classrooms become hubs for developing critical thinking skills.

Communication: Language classrooms naturally lend themselves to developing communication skills. Teachers can incorporate activities that promote effective oral and written communication. This can include group discussions, role-plays, presentations, and collaborative writing projects. Students can also engage in authentic communication through tasks such as interviews, debates, or writing letters or emails to native speakers. By providing ample opportunities for meaningful communication, language classrooms become spaces where students refine their communication skills.

Collaboration: Collaboration is a vital skill in the 21st century, and language classrooms offer an ideal setting for fostering collaborative learning. Teachers can design cooperative activities such as group projects, problem-solving tasks, or language exchanges. Through collaborative tasks, students develop teamwork, negotiation, and leadership skills. They learn to respect diverse perspectives, actively listen, and work towards shared goals, preparing them for collaborative endeavours beyond the classroom.

Creativity: Language learning can be a catalyst for nurturing creativity. Teachers can encourage students to think creatively by incorporating tasks that require imagination, originality, and self-expression. This can involve creative writing, storytelling, or artistic projects where students create visual representations or performances related to language content. By embracing creativity, language classrooms become spaces where students can express their ideas, experiment with language, and develop innovative approaches to communication.

Digital Literacy: In today's digital age, digital literacy is crucial for effective communication and information processing. Language classrooms can integrate digital tools and resources to enhance language learning experiences. This can include using online language learning platforms, multimedia content, interactive websites, or language learning apps. Teachers can guide students in evaluating online sources, creating multimedia presentations, and leveraging technology for collaborative projects. By promoting digital literacy, language classrooms equip students with essential skills for navigating the digital world.

To effectively integrate 21st-century skills, teachers can employ pedagogical strategies that focus on student-centred learning, inquiry-based approaches, and project-based learning. They can provide opportunities for reflection, self-assessment, and goal-setting, allowing students to take ownership of their learning and develop metacognitive skills. Teachers can also provide constructive feedback and encourage students to reflect on their learning process and the development of 21st-century skills.

Additionally, integrating authentic materials, real-world contexts, and cross-curricular connections can make language learning more relevant and meaningful. This can involve incorporating current events, cultural artifacts, or multimedia content that reflect real-world language use and diverse perspectives.

Integrating technology in language classrooms can enhance the development of 21st-century skills. For example, students can use online collaboration tools, digital storytelling platforms, or multimedia creation tools to engage in collaborative and creative tasks. Teachers can guide students in navigating digital resources responsibly and critically, promoting digital literacy.

Integration of 21st-century skills in language classrooms requires intentional planning, innovative approaches, and a student-centred focus. By nurturing these skills alongside language proficiency, educators prepare students for success in an ever-evolving global landscape, equipping them with the skills needed to communicate effectively, think critically, collaborate productively, and navigate the digital realm with confidence.

For example, AI algorithms can generate individualized language exercises, recommend targeted practice materials, and provide instant feedback on pronunciation and grammar.

Virtual and Augmented Reality: Virtual and augmented reality technologies offer immersive language learning experiences. Students can virtually travel to different countries and engage in simulated conversations with native speakers. They can explore virtual language environments, such as ordering food at a restaurant or negotiating in a business meeting. These technologies enhance cultural understanding and provide authentic language practice opportunities.

Gamification and Social Learning: Gamification techniques can make language learning more engaging and motivating. Language learning apps can incorporate game elements, such as leaderboards, badges, and rewards, to encourage progress and competition. Additionally, social learning platforms can facilitate language exchanges and collaborative projects among students worldwide, fostering a sense of community and cultural exchange.

Content-Based Instruction: Language learning will continue to integrate with other subjects and real-world contexts. Content-based instruction involves teaching language skills through the exploration of meaningful topics, such as history, science, or current events. This approach promotes interdisciplinary learning and enhances language acquisition by connecting language skills to authentic and relevant contexts.

Blended Learning and Flipped Classrooms: Blended learning, combining traditional classroom instruction with online resources, will become more prevalent. Teachers can assign online language lessons and interactive activities for students to complete outside of class, allowing for more in-depth and interactive discussions during classroom time. Flipped classrooms, where students access instructional materials before class and engage in collaborative activities during class, will also gain popularity, fostering student-centered learning and language practice.

Multimodal Learning: Language learning will increasingly incorporate multiple modes of communication, including visual, auditory, and kinesthetic elements. Teachers can integrate multimedia resources, such as videos, podcasts, and interactive presentations, to engage learners and cater to different learning styles. Additionally, hands-on activities, role-plays, and real-life simulations can provide experiential learning opportunities and enhance language proficiency.

Emphasis on Cultural Competence: In an increasingly globalized world, cultural competence will play a vital role in language learning. Future classrooms will prioritize developing intercultural communication skills and fostering understanding and respect for diverse cultures. Virtual cultural exchanges, guest speakers, and authentic cultural materials will be integrated into language curricula to promote cross-cultural understanding and proficiency.

Lifelong Learning and Microcredentials: Language learning will extend beyond the classroom, and individuals will engage in lifelong learning to enhance their language skills for personal and professional growth. Microcredentials, such as digital badges or certificates, will gain recognition as individuals demonstrate specific language competencies through online courses, language assessments, and project-based portfolios.

To prepare for the future of language learning, educators can adopt the following strategies:

a) Embrace technology: Incorporate language learning apps, online platforms, and digital resources to enhance instruction and provide personalized learning experiences.

b) Foster a learner-centred approach: Encourage students to take ownership of their language learning, set goals, and engage in self-assessment and reflection.

c) Cultivate cultural awareness: Integrate cultural components into language instruction, promote cultural exchange, and develop students' intercultural competence.

d) Provide authentic and meaningful language experiences: Create opportunities for real-world language use, such as collaborative projects, community engagement, and online language exchanges.

e) Continuously update teaching methodologies: Stay informed about the latest research and trends in language pedagogy, attend professional development workshops, and collaborate with colleagues to share innovative practices.

By embracing these strategies and staying adaptable to emerging technologies and pedagogical approaches, educators can prepare students for the future of language learning, equipping them with the language skills and 21st-century competencies needed to thrive in an increasingly interconnected world.

The process of implementing project-based learning for the integration of language learning and 21st century skills can be divided into several steps:

Topic Selection: Begin by selecting a relevant and engaging topic that aligns with the language learning objectives and allows for the application of 21st century skills. For example, the topic could be creating a marketing campaign for a new environmentally friendly product.

Project Design: Design the project to provide students with authentic and meaningful tasks. Provide clear guidelines and expectations for the project, including the specific language skills to be developed and the 21st century skills to be applied. For instance, students will need to conduct market research, analyse target audiences, and develop persuasive communication materials.

Preparing Resources: Gather and prepare the necessary resources and materials for the project. This may include access to relevant research materials, graphic design software, presentation tools, and other resources that students may need to complete their tasks.

Group Formation: Form groups or teams to encourage collaboration and teamwork. Consider mixing students with different language proficiency levels to promote peer learning and support. Assign specific roles and responsibilities within each group, such as project manager, researcher, writer, designer, and presenter.

Planning and Research: Provide students with time to plan and conduct research related to their marketing campaign. This includes identifying target audiences, analysing competitors, and understanding market trends. Students can use both online and offline resources to gather information.

Language Skills Development: Throughout the project, focus on language skills development. Provide language input and support as needed, such as vocabulary related to marketing and persuasive writing, grammar structures for effective communication, and strategies for presenting ideas clearly.

Collaboration and Communication: Encourage regular collaboration and communication among group members. Set milestones and deadlines to ensure progress and provide opportunities for students to give and receive feedback on their work. This fosters teamwork, effective communication, and collaboration skills.

Creative Output: Allow students to showcase their creativity by designing persuasive advertisements, creating visual materials, and developing compelling presentations. Encourage the use of multimedia tools and technology to enhance the visual appeal and effectiveness of their marketing campaign.

Presentation and Reflection: Provide students with opportunities to present their marketing campaigns to the class or a wider audience. This allows them to practice their speaking skills, receive feedback, and reflect on their learning process. Encourage students to reflect on their experience, identify challenges, and discuss how they applied 21st century skills throughout the project.

To innovate the process and enhance the integration of language learning and 21st century skills, consider the following approaches:

Authentic Audience: Provide students with opportunities to present their marketing campaigns to a real audience beyond the classroom, such as local businesses, community organizations, or online platforms. This adds authenticity to the project and gives students a sense of purpose and impact.

Technology Integration: Encourage the use of digital tools and platforms for conducting research, designing visual materials, and creating multimedia presentations. This not only enhances students' digital literacy but also allows for more dynamic and interactive project outcomes.

Global Connections: Facilitate connections and collaborations with students from different cultural backgrounds or schools in other countries. Through virtual exchanges or collaborative projects, students can develop cross-cultural communication skills and gain a broader perspective on global marketing strategies.

Reflection and Self-Assessment: Incorporate regular opportunities for students to reflect on their learning, assess their progress, and set goals for improvement. This promotes metacognition and self-directed learning, allowing students to take ownership of their language development and 21st century skill acquisition.

By continuously innovating project-based learning approaches, teachers can maximize the integration of language learning and 21st century skills, creating engaging and impactful learning experiences for students.

Exemplifying Integration in Language Classrooms

Collaborative Storytelling

Through Collaborative Storytelling, students not only improve their language skills but also hone their collaboration, creativity, and digital literacy abilities. Designing a collaborative storytelling project involves several key steps to ensure its success. Here is an overview of the process:

Defining the Objective: Start by clarifying the objective of the collaborative storytelling project. Determine what specific language skills and 21st-century skills you want students to develop, such as vocabulary expansion, narrative structure, collaboration, and creativity.

Selecting the Format: Decide on the format for the collaborative storytelling project. It could be a written story, a digital storybook, a multimedia presentation, or even a video production. Consider the available resources, technology tools, and the interests of your students.

Forming Collaborative Groups: Divide the students into small groups to foster collaboration. Aim for diverse groups that bring together students with different language abilities, strengths, and interests. Assign roles within each group, such as a writer, editor, illustrator, and technology specialist, to encourage teamwork.

Choosing a Theme or Prompt: Provide a theme or prompt to guide the storytelling process. It could be a specific topic, a picture, a sentence, or even a word. The theme should be open-ended enough to allow for creativity and different interpretations.

Brainstorming and Planning: Encourage students to brainstorm ideas, characters, settings, and plot elements related to the theme or prompt. Guide them in developing a storyline and creating a storyboard or outline. Emphasize the importance of collaboration, active listening, and respecting each other's ideas during this stage.

Writing and Editing: In this phase, each group begins writing their part of the story based on the agreed-upon outline. Encourage students to use descriptive language, dialogue, and narrative techniques to engage the readers. As they progress, emphasize the importance of peer editing and providing constructive feedback to improve the story's coherence and quality.

Incorporating Digital Elements: If the project involves digital storytelling, students can use various digital tools and resources to enhance their story. This may include adding visuals, audio recordings, animations, or interactive elements. Provide guidance on the use of digital tools and encourage students to explore different options based on their creativity and technical abilities.

Rehearsing and Presenting: Once the story is complete, students can rehearse and practice presenting their collaborative storytelling project to the class or a wider audience. This allows them to further develop their oral communication skills and build confidence in sharing their work.

Reflection and Evaluation: Provide opportunities for students to reflect on the collaborative storytelling project. Encourage them to discuss what they have learned, the challenges they faced, and the skills they developed. Additionally, consider implementing self-assessment and peer assessment activities to evaluate individual and group contributions.

By following these steps, teachers can design a collaborative storytelling project that promotes language development, collaboration, creativity, and digital literacy. This project-based approach engages students actively, encourages teamwork, and fosters the acquisition of 21st-century skills in an authentic and meaningful context.

Critical Thinking in Reading Comprehension

This strategy is all about thought-provoking questions, analysis tasks, and discussions where students develop their language proficiency while also enhancing their critical thinking skills. Selecting Suitable Texts: Choose reading materials that are pertinent, captivating, and suitable for the students' language proficiency level. Opt for fiction or non-fiction texts, such as short stories, articles, essays, or excerpts from novels. Let us have a look at a flexible plan for a language classroom process to ignite critical thinking in a classroom that focusses on reading comprehension

Pre-Reading Stage:

Introduce the reading passage and activate prior knowledge by asking open-ended questions related to the topic.

Example: "What do you already know about this topic? What questions do you have?"

Establishing Reading Purpose:

Clearly define the purpose of reading and set specific goals for comprehension.

Example: "As you read, pay attention to the author's main argument and supporting evidence."

Pre-Reading Critical Thinking Skills:

Teach and model critical thinking skills such as making predictions, formulating questions, and activating schema.

Example: "Based on the title and the introductory paragraph, what do you predict the author's viewpoint might be?"

During-Reading Strategies:

Encourage active reading by using strategies like annotation, highlighting key points, and note-taking.

Example: "As you read, underline any statements that support the author's argument."

Analysing Author's Argument:

Engage students in evaluating the author's claims, evidence, and reasoning.

Example: "What evidence does the author provide to support their argument? How strong is the evidence?"

Identifying Bias and Assumptions:

Guide students to recognize any biases or assumptions present in the text and analyse their impact on the author's argument.

Example: "What values or beliefs might the author hold that could influence their perspective?"

Evaluating Credibility and Validity:

Encourage students to assess the credibility of the sources cited and evaluate the validity of the evidence provided.

Example: "Do you think the sources the author cites are reliable? Why or why not?"

Reflecting on Personal Perspectives:

Prompt students to reflect on their own beliefs, opinions, and biases in relation to the text.

Example: "How does this text challenge or confirm your own perspective on the topic?"

Post-Reading Discussion and Analysis:

Facilitate meaningful discussions where students share their interpretations, support their claims with evidence, and challenge each other's ideas.

Example: "Present your argument and provide evidence from the text to support your viewpoint. Respond to your classmates' perspectives with thoughtful questions."

Reflective Writing or Response:

Assign reflective writing tasks that require students to synthesize their understanding, critically evaluate the text, and provide their own analysis.

Example: "Write a response discussing the strengths and weaknesses of the author's argument. Include your own perspective and support it with evidence."

These strategies should be adapted based on the students' proficiency level and the complexity of the reading material. By incorporating these critical thinking strategies, teachers can encourage students to analyse texts, question assumptions, evaluate evidence, and develop their overall critical thinking skills in the context of reading comprehension. By integrating thought-provoking inquiries, analytical tasks, and group dialogues into the reading comprehension lesson, teachers can cultivate students' critical thinking skills while developing their language proficiency. This process entails guiding students to analyse, interpret, evaluate, and synthesize information from the text, ultimately enhancing their capacity for critical thinking and fostering deeper engagement with the material.

Virtual Exchange Collaboration

Through online platforms and video conferencing tools, students collaborate on various language-learning activities, including discussions, debates, and project-based tasks. This virtual exchange not only improve the language proficiency of learners but also foster cultural understanding, communication skills, and global citizenship.

Partnering with a Distant Classroom: Connect with a language classroom in a different geographic location or cultural context through virtual exchange platforms or educational networks. Collaborate with the partner teacher to design joint activities and establish communication channels.

Intercultural Introductions: Begin the virtual exchange by having students introduce themselves to their partner classmates through video introductions or written profiles. Encourage students to share information about their culture, interests, and language-learning goals.

Online Discussions: Organize online discussions on specific topics or themes related to language learning or cultural aspects. Students can use video conferencing tools or online discussion boards to share their thoughts, ask questions, and engage in meaningful conversations with their peers from the partner classroom. Encourage active listening, respectful dialogue, and the exchange of different perspectives.

Collaborative Projects: Assign collaborative projects that require students from both classrooms to work together towards a common goal. This could involve creating a joint presentation, conducting research on a shared topic, or collaborating on a digital storytelling project. Encourage students to use their language skills, collaborate effectively, and employ digital tools for communication and content creation.

Language Exchanges: Facilitate language exchanges where students take turns practicing their target language with their partner classmates. This can be done through video chats, voice messages, or written exchanges. Encourage students to provide feedback on each other's language use and support each other's language development.

Cultural Exchanges: Promote cultural understanding by organizing activities that allow students to share aspects of their culture with their partner classmates. This could include virtual tours of their hometown, presentations on cultural traditions, or sharing examples of music, art, or literature from their culture. Encourage students to ask questions, compare cultural practices, and appreciate the diversity of perspectives.

Reflective Journals: Have students maintain reflective journals to document their experiences and reflections throughout the virtual exchange collaboration. Encourage them to reflect on their language development, intercultural interactions, and the skills they have acquired during the process. This helps students deepen their understanding of the learning experience and promotes metacognition.

Teacher Facilitation and Feedback: As the teacher, provide guidance, support, and feedback to students throughout the virtual exchange collaboration. Offer prompts for discussions, facilitate online interactions, and monitor the progress of collaborative projects. Provide constructive feedback on language use, intercultural communication, and the demonstration of 21st century skills.

By engaging in virtual exchange collaboration, students not only enhance their language skills but also develop important 21st century skills such as communication, collaboration, cultural understanding, and global citizenship. Through interactions with their peers from different backgrounds, students gain a broader perspective, improve their intercultural competence, and develop valuable skills for an interconnected world.

Gamification for Language Acquisition

Language teachers can incorporate gamification elements into the curriculum to enhance language acquisition and 21st-century skill development. By using language learning apps, educational games, and online interactive platforms, students engage in immersive and interactive experiences that promote language proficiency, problem-solving, creativity, and digital literacy.

Word Building Challenge:

Students take turns adding letters to create new words. Start with a single letter, and each student adds a letter to create a new word. The word should be valid and related to the topic or category. Encourage students to explain the meaning of their words and use them in contexts.

It promotes vocabulary expansion, spelling, and word formation skills.

Story Starters:

Students are given a sentence or a prompt to start a story, and they continue building the narrative collaboratively. Provide each student with a sentence or a prompt to start a story. Students take turns adding paragraphs or sections to continue the story. Encourage them to incorporate new vocabulary and grammatical structures.

Story Starters stimulate creative thinking, storytelling skills, and cooperative writing.

Language Learning Apps:

Duolingo, Memrise, Babbel, and similar apps offer interactive lessons, vocabulary practice, and language exercises.

They provide personalized learning experiences, adaptive feedback, and progress tracking.

These apps can be used by students independently or integrated into classroom activities, allowing learners to practice language skills at their own pace.

Virtual Reality (VR) Language Experiences:

Immersive VR experiences allow students to virtually travel to different countries and engage in language learning scenarios. They provide authentic language exposure, cultural immersion, and the opportunity to practice real-world language skills. Teachers can use VR headsets or online VR platforms to create virtual language trips, language simulations, or virtual language exchanges.

Online Language Games:

Platforms like Kahoot, Quizlet, and Quizizz offer language learning games that reinforce vocabulary, grammar, and language usage. They promote competition, engagement, and active learning through interactive quizzes, word challenges, and language puzzles. Teachers can create custom quizzes or use existing game templates to review language concepts and assess student understanding.

Gamified Language Learning Platforms:

FluentU, Rosetta Stone, and Lingoda provide gamified language learning experiences with interactive lessons, videos, and exercises. They incorporate gamification elements such as points, badges, and levels to motivate and engage learners. These platforms offer a structured curriculum, progress tracking, and opportunities for self-paced learning and skill development.

Online Language Exchange Platforms:

Websites like HelloTalk, Tandem, and Italki connect language learners worldwide for virtual language exchange and practice. Students can interact with native speakers, engage in conversations, and receive feedback on their language skills. Language exchange platforms foster communication skills, cultural understanding, and global connections, promoting the development of 21st-century skills.

These digital language learning games and platforms serve the cause of language acquisition and 21st-century skills enhancement by providing engaging and interactive experiences. They offer opportunities for authentic language practice, real-time feedback, and personalized learning. Through gamification elements, they motivate students to actively participate and progress in their language learning journey. These games can be undertaken in classrooms through computer labs, mobile devices, or online learning platforms, enabling teachers to monitor student progress, provide guidance, and facilitate meaningful discussions and collaboration.

Social Media Language Learning Community

Social Media Language Learning Community highlights a language classroom that utilize social media platforms to create a language learning community. Students interact with native speakers, share language-related content, and engage in discussions and collaborative projects. This approach enhances their language skills, digital literacy, communication, and cross-cultural understanding.

Language Learning Groups on social media:

Create dedicated language learning groups on platforms like Facebook, WhatsApp, or Slack. Encourage students to join these groups to interact with peers, share resources, ask questions, and practice language skills. Teachers can facilitate discussions, provide language prompts, and monitor student participation.

Language Challenges and Hashtags:

Design language challenges or assignments that require students to create and share language-related content using specific hashtags. For example, students can post short videos, pictures, or written compositions on Instagram or Twitter using a designated hashtag.

Language Challenges and Hashtags encourage students to actively use the language, receive feedback from peers and teachers, and build a sense of community.

Collaborative Projects and Blogs:

Assign collaborative projects where students work in groups to create and maintain language learning blogs or websites. Students can share their writings, reflections, or multimedia content related to language learning and cultural topics.

This fosters collaboration, creativity, digital literacy, and the development of writing and presentation skills.

Virtual Language Exchanges:

Facilitate virtual language exchange partnerships with students from other locations or with native speakers. Students can communicate via social media platforms or video conferencing tools to practice speaking and listening skills.

Virtual Language Exchanges provide authentic language practice, cultural exchange, and enhances communication and intercultural competence.

Authentic Materials and Discussions:

Encourage students to follow social media accounts or join groups related to their target language and culture. Students can engage with authentic materials such as articles, videos, or podcasts and participate in discussions related to those topics.

Authentic Materials and Discussions develop reading and listening skills, critical thinking, and cross-cultural understanding.

These strategies serve the cause of language acquisition and 21st-century skills enhancement by providing authentic and engaging language practice in real-world contexts. Social media platforms allow students to interact with a larger community of language learners and native speakers, fostering communication skills, cultural understanding, and digital literacy. Through collaborative projects, language challenges, and virtual exchanges, students develop collaboration, creativity, and adaptability. These platforms can be usefully undertaken in classrooms by establishing guidelines for responsible and respectful use, setting clear learning objectives, and providing support and guidance throughout the process. Teachers can monitor student activities, provide feedback, and facilitate meaningful discussions that encourage critical thinking and reflection.

Online Language Portfolio and Reflections

In Online Language Portfolio and Reflections, students maintain online language portfolios to document their language learning journey. They regularly upload samples of their work, reflections on their progress, and feedback from peers and teachers. This approach foster self-assessment, metacognition, and critical thinking as students actively reflected on their language development, set goals, and tracked their growth.

Digital Portfolio Platforms:

Utilize online platforms or learning management systems that allow students to create and maintain their digital language portfolios. Platforms like Seesaw, Google Sites, or WordPress can be used to upload and organize samples of student work, such as written compositions, audio recordings, or multimedia projects. Students can regularly update their portfolios to showcase their language skills development over time.

Reflective Journals or Blogs:

Encourage students to maintain reflective journals or blogs where they can write about their language learning experiences, challenges, and progress. Students can reflect on their language goals, strategies they have used, and areas for improvement.

This promotes metacognition, critical thinking, and self-assessment as students analyse their strengths and weaknesses in language learning.

Peer Feedback and Collaboration:

Incorporate peer feedback as a component of the portfolio process. Students can provide constructive feedback on their peers' work and reflect on the feedback they receive.This promotes communication skills, collaboration, and a growth mindset as students learn from each other's perspectives.

Goal Setting and Progress Tracking:

Guide students in setting specific, measurable, attainable, relevant, and time-bound (SMART) language goals. Students can regularly update their portfolios with reflections on their progress towards these goals. This helps students take ownership of their learning, develop self-regulation skills, and track their growth over time.

Teacher Feedback and Assessment:

Teachers can provide regular feedback on students' portfolio entries to guide their language development. This feedback can focus on language accuracy, organization, critical thinking, and creativity. Teachers can also conduct periodic assessments based on the portfolio entries to evaluate language proficiency and provide targeted support.

Online language portfolios and reflections serve the cause of language acquisition and 21st-century skills enhancement by promoting self-assessment, metacognition, critical thinking, and goal setting. They provide students with a platform to actively reflect on their language learning journey, identify areas for improvement, and take ownership of their progress. By documenting their work and receiving feedback from peers and teachers, students develop communication skills, collaboration, and the ability to think critically about their language development. These portfolios can be usefully undertaken in classrooms by providing clear guidelines and expectations for portfolio entries, dedicating time for reflection and goal setting, and facilitating opportunities for peer feedback and discussion. Teachers can review and provide feedback on students' portfolios, track their progress, and use the portfolios as evidence of language proficiency growth during assessments and parent-teacher conferences.

Artificial Intelligence Language Tutoring

Artificial Intelligence Language Tutoring explores the integration of artificial intelligence (AI) language tutoring tools in the classroom. Students interact with AI-powered chatbots or virtual language tutors, which provide personalized feedback, language practice, and support. This approach not only enhance their language proficiency but also promoted adaptability, digital literacy, and self-directed learning skills.

The integration of artificial intelligence (AI) language tutoring tools in the classroom can revolutionize language learning and promote the development of 21st-century skills. Here are some examples and futuristic possibilities that illustrate the potential of AI language tutoring:

Intelligent Chatbots:

Students can interact with AI-powered chatbots that simulate conversations in the target language. Chatbots can provide personalized feedback on grammar, vocabulary, pronunciation, and language usage in real-time. For example, a student can engage in a conversation with a chatbot about a specific topic and receive instant feedback on their language accuracy and fluency.

Adaptive Learning Platforms:

AI-based language learning platforms can analyse students' strengths, weaknesses, and learning styles to tailor instruction. The platforms can adapt the curriculum and provide customized language exercises and activities based on individual needs. For instance, a platform can identify that a student struggles with verb tenses and provide targeted exercises to reinforce that specific area of language acquisition.

Virtual Language Tutors:

Advanced AI technologies can simulate virtual language tutors that guide students through interactive lessons and provide personalized instruction. Virtual tutors can offer instant feedback, scaffold learning, and adjust the difficulty level based on student performance. They can engage students in dialogues, role-plays, and interactive exercises to practice language skills in a dynamic and engaging way.

Natural Language Processing (NLP) Technologies:

NLP technologies can analyse students' written or spoken responses and provide detailed feedback on grammar, vocabulary, and sentence structure. These technologies can identify common errors, suggest corrections, and offer explanations to help students improve their language accuracy. As an example, a student can write an essay and receive instant feedback on grammatical errors and suggestions for improvement from the AI-powered system.

Virtual Reality (VR) Language Immersion:

Future advancements in VR technology can create immersive language learning environments where students can practice language skills in realistic scenarios. Students can engage in virtual conversations with AI-powered characters, navigate virtual cultural settings, and experience authentic language interactions. For instance, students can explore a virtual marketplace where they must negotiate in the target language with AI-generated shopkeepers.

These innovative possibilities demonstrate how AI language tutoring can enhance language proficiency while promoting adaptability, digital literacy, and self-directed learning skills. By interacting with AI-powered tools, students receive personalized feedback, engage in authentic language practice, and develop their ability to adapt to different language contexts. AI language tutors can offer continuous support, adapt the curriculum to individual needs, and provide immediate feedback, enabling students to learn at their own pace and take ownership of their language learning journey. Furthermore, the integration of AI technology opens up new avenues for immersive language experiences, personalized instruction, and expanded access to language learning resources.

Global Collaboration through Online Projects

In Global Collaboration through Online Projects, students from different parts of the world collaborate on online projects to enhance their language skills and foster cultural understanding. Through video conferencing, collaborative document editing, and online platforms, students engaged in cross-cultural exchanges, project-based learning, and digital collaboration. This approach fosters language proficiency, communication skills, teamwork, and digital literacy.

Global collaboration through online projects provides students with valuable opportunities to enhance their language skills while fostering cross-cultural understanding and developing 21st-century skills. Here are some examples of strategies that can be useful in the language classroom to support language acquisition and 21st-century skills enhancement through global collaboration:

Virtual Pen Pals:

Students can connect with peers from different countries through virtual pen pal exchanges. They can exchange emails, video messages, or participate in online forums to practice their language skills and learn about different cultures. For example, students can engage in regular written exchanges with their pen pals, discussing various topics and sharing insights into their respective cultures.

Collaborative Projects:

Students from different parts of the world can collaborate on projects that require language use and cultural exploration.

They can work together on research projects, presentations, or multimedia creations, using online collaboration tools and platforms.

For instance, students can collaborate on creating a digital magazine about a specific cultural theme, each contributing articles, visuals, and multimedia content.

Video Conferencing:

Students can engage in video conferencing sessions with peers from other countries to practice conversational skills and cultural exchanges.

They can engage in discussions, debates, or role-plays, focusing on topics of mutual interest.

For example, students can participate in a video conference where they discuss global environmental issues, sharing perspectives and proposing solutions.

Online Language Exchanges:

Students can participate in online language exchange programs where they connect with native speakers of the target language.

They can engage in language conversations, language sharing, and cultural exchanges through voice or video calls.

For instance, students can schedule regular language exchange sessions with native speakers, practicing their speaking and listening skills while learning about different cultures.

Global Collaborative Writing:

Students from different countries can collaborate on writing projects, such as storytelling or creating a shared document.

They can take turns adding paragraphs or sections to a story, building upon each other's ideas and language skills.

For example, students can collectively write a digital book, each contributing chapters or sections in their respective languages, showcasing their language proficiency and creativity.

Global collaboration through online projects serves the cause by providing authentic language practice opportunities, fostering cultural understanding, and developing 21st-century skills. Students engage in meaningful interactions with peers from different cultural backgrounds, expanding their worldview and developing intercultural communication skills. They learn to collaborate effectively, navigate digital platforms, and develop their digital literacy. Additionally, global collaboration promotes curiosity, empathy, and adaptability as students encounter diverse perspectives and cultural practices. By undertaking these strategies, language classrooms become dynamic and inclusive spaces where students acquire language skills while developing the essential skills needed to thrive in the 21st century.

Podcasting for Language Expression

This case study explores a language classroom where students created podcasts as a means of expressing their ideas, thoughts, and opinions. By researching, scripting, and recording their own podcasts, students honed their language skills, creativity, critical thinking, and digital communication abilities. The chapter provides insights into the podcasting process, tools used, and the positive impact on student engagement and language development.

Virtual Reality Cultural Immersion

In this case study, students experienced virtual reality (VR) simulations to immerse themselves in different cultures and practice language skills in authentic contexts. Using VR headsets and interactive VR platforms, students engaged in virtual travel, cultural activities, and language interactions. This immersive approach enhanced their language proficiency, cultural competence, empathy, and digital literacy.

Online Language Learning Communities

This case study highlights the use of online language learning communities, such as forums, social media groups, and language exchange platforms. Students joined these communities to connect with native speakers, engage in language practice, and receive feedback on their language skills. This approach fostered language acquisition, communication skills, cultural understanding, and digital citizenship.

Data-driven Language Learning Analytics

In this case study, language teachers utilized data-driven language learning analytics tools to monitor students' progress, identify areas of improvement, and provide personalized feedback. By analysing language learners' performance data, teachers gained insights into individual learning patterns, tailored instruction, and facilitated self-directed learning. This approach promoted self-reflection, goal-setting, critical thinking, and data literacy skills.

Assessment of 21st Century Skills in Language Learning

The integration of 21st century skills in language classrooms is a transformative approach that empowers learners to become effective communicators, critical thinkers, and adaptable problem solvers. Through pedagogical strategies and real-life case studies, this chapter highlights the practical implementation of these skills in language education. By embracing the fusion of language learning and 21st century skills, educators can prepare learners to thrive in an interconnected and rapidly evolving world.

Assessing 21st-century skills in language learning is crucial to measure students' progress and provide meaningful feedback. Traditional assessment methods often focus on language proficiency alone, but incorporating assessment strategies for 21st-century skills ensures a comprehensive evaluation of students' abilities. Here's an explanation of the processes and strategies for assessing 21st-century skills in language learning, along with the role of technology in facilitating assessment, recording, and reporting:

Clear Learning Outcomes: Begin by establishing clear learning outcomes for 21st-century skills aligned with language learning objectives. These outcomes should specify the skills to be assessed and the criteria for proficiency levels. For example, if assessing communication skills, specify the expected level of fluency, accuracy, and coherence in spoken or written communication.

Authentic Assessments:

Design authentic assessments that mirror real-world scenarios where students can apply 21st-century skills. This can include performance-based tasks such as oral presentations, debates, collaborative projects, or problem-solving activities. Authentic assessments provide a holistic view of students' skills and encourage them to demonstrate their abilities in meaningful contexts.

Rubrics and Criteria:

Develop clear rubrics and assessment criteria to evaluate 21st-century skills. Rubrics provide a framework for assessing different aspects of the skills and enable consistent and fair evaluation. Clearly define the criteria for each proficiency level, ensuring that they align with the learning outcomes and provide students with a clear understanding of expectations.

Technology-Enhanced Assessments:

Technology can play a significant role in facilitating the assessment of 21st-century skills. It offers various tools and platforms for creating interactive assessments, collecting data, and providing immediate feedback. For example, online platforms can be used for collaborative writing tasks, video conferencing for oral communication assessments, or digital portfolios for students to showcase their projects and reflections.

Multimedia Projects: Incorporate multimedia projects as assessment tasks where students can showcase their creativity, digital literacy, and communication skills. This can involve creating videos, podcasts, or multimedia presentations that require students to research, analyse, synthesize information, and present their findings in engaging and innovative ways.

Peer and Self-Assessment: Encourage peer and self-assessment as valuable assessment strategies for 21st-century skills. Students can evaluate their own progress, reflect on their strengths and areas for improvement, and set goals for further development. Peer assessment allows students to provide feedback to their peers, fostering collaboration, critical thinking, and communication skills.

Data Collection and Analysis:

Use technology tools to collect and analyse data from assessments. Online platforms or software can facilitate data collection, allowing teachers to gather quantitative and qualitative data on students' performance in 21st-century skills. This data can then be analysed to identify trends, strengths, and areas for improvement.

Feedback and Reflection:

Provide constructive feedback to students on their performance in 21st-century skills. Focus on specific strengths and areas for growth, highlighting examples and suggesting strategies for improvement. Encourage students to reflect on their own performance, set goals, and engage in self-directed learning.

Digital Portfolios:

Implement digital portfolios where students can compile evidence of their work, including projects, reflections, and assessments related to 21st-century skills. Digital portfolios allow students to showcase their growth over time and provide a comprehensive view of their abilities. Technology platforms can facilitate the creation, organization, and sharing of digital portfolios.

Reporting and Parent Communication:

Technology can assist in recording and reporting assessment results. Online platforms or learning management systems can generate reports that provide a comprehensive overview of students' proficiency in 21st-century skills. These reports can be shared with parents to communicate students' progress and strengths in different skill areas.

Canto Seven
Numeric Enigmas: Unveiling the Mathematical Symphony

In this chapter, we examine the integration of 21st-century skills in mathematics education, focusing on the use of advanced technology and pedagogical approaches that promote critical thinking, problem-solving, collaboration, creativity, and digital literacy. As our world becomes increasingly digital, it is essential for math classrooms to embrace these advancements and prepare students for the challenges and opportunities of the future. We will explore practical strategies and activities that educators can employ to foster 21st-century skills in their math classrooms, creating engaging and innovative learning environments.

In the 21st century, education has shifted its focus from rote memorization and procedural knowledge to a more holistic approach that encompasses a wide range of skills required for success in the digital age. Mathematics education, in particular, plays a crucial role in developing these skills, as it provides a foundation for logical thinking, problem-solving, and analytical reasoning. This section aims to define and explore the key 21st-century skills that are relevant in mathematics education, highlighting their significance and providing examples of how they can be integrated into the classroom.

Critical Thinking:

Critical thinking is a fundamental skill necessary for students to navigate the complexities of the modern world. In mathematics education, critical thinking involves analysing and evaluating information, making connections between concepts, and drawing logical conclusions. For example, when solving a complex mathematical problem, students need to analyse the problem, break it down into smaller parts, consider alternative approaches, and evaluate the reasonableness of their solutions.

Collaboration:

Collaboration is an essential skill for the 21st century, as it enables individuals to work effectively in teams and leverage diverse perspectives to solve complex problems. In mathematics education, collaboration can take various forms, such as group projects, peer tutoring, or online discussions. For instance, students can work together to solve challenging math problems, explaining their reasoning and strategies to one another, and collectively arrive at a solution. This not only enhances their mathematical understanding but also fosters communication, teamwork, and interpersonal skills.

Creativity:

Creativity involves thinking outside the box, generating innovative ideas, and finding unique solutions to problems. In mathematics education, creativity can be nurtured through open-ended problem-solving tasks that allow students to explore multiple approaches and develop their own strategies. For example, students can be presented with a real-world problem and asked to devise creative ways to apply mathematical concepts to solve it. This encourages them to think creatively, explore alternative solutions, and see mathematics as a tool for practical applications.

Problem-Solving:

Problem-solving is a core skill that lies at the heart of mathematics education. It involves identifying problems, formulating strategies, and applying mathematical concepts and procedures to find solutions. In the 21st century, problem-solving extends beyond traditional textbook exercises and requires students to tackle real-world problems that are open-ended and multifaceted. For instance, students can be presented with a complex mathematical puzzle or a real-life scenario that requires them to analyse data, make predictions, and formulate strategies to solve the problem. This develops their critical thinking, analytical skills, and perseverance.

Digital Literacy:

Digital literacy refers to the ability to use digital technologies effectively and critically to access, evaluate, and create information. In mathematics education, digital literacy can be integrated by incorporating technology tools, such as graphing calculators, interactive simulations, and data analysis software. For example, students can use spreadsheets to analyse and visualize data, explore mathematical patterns using online graphing tools, or engage in online collaborative platforms to share and discuss mathematical ideas. This not only enhances their mathematical skills but also equips them with the digital literacy skills necessary for the 21st century.

Integrating 21st-century skills into mathematics education is essential for preparing students to thrive in a digital and interconnected world. By fostering critical thinking, collaboration, creativity, problem-solving, and digital literacy within the mathematics classroom, educators can empower students with the skills they need for success in their academic and professional lives. Through engaging activities, real-world applications, and the integration of technology, mathematics education can become a catalyst for the development of these essential 21st-century skills.

The Importance of Integrating 21st-Century Skills in Math:

In the 21st century, the world is rapidly evolving, driven by technological advancements and an increasing demand for skills that go beyond traditional knowledge acquisition. As a result, there is a growing recognition of the importance of integrating 21st-century skills into education to prepare students for success in an increasingly complex and interconnected world. Mathematics, as a discipline, is uniquely positioned to foster these skills, providing students with the tools to think critically, solve problems, and make connections to real-world applications. This section highlights the benefits of integrating 21st-century skills in math classrooms, emphasizing the positive impact on problem-solving abilities, critical thinking, and real-world application.

Enhanced Problem-Solving Abilities:

Integrating 21st-century skills in math education significantly enhances students' problem-solving abilities. Traditional math education often focuses on rote memorization and procedural knowledge, which can limit students' ability to apply mathematical concepts in novel and complex situations. By incorporating 21st-century skills such as critical thinking, collaboration, and creativity, math classrooms become environments that encourage students to think critically, analyse problems from multiple perspectives, and devise innovative solutions. This approach equips students with the skills necessary to tackle real-world challenges that require flexible thinking and creative problem-solving.

Developing Critical Thinking Skills:

Mathematics provides an ideal platform for developing critical thinking skills. By integrating 21st-century skills in math classrooms, students are encouraged to question, analyse, and evaluate information, leading to a deeper understanding of mathematical concepts. Critical thinking enables students to make logical connections, identify patterns, and evaluate the validity of mathematical arguments. It also cultivates a mindset of inquiry and curiosity, empowering students to explore and discover mathematical concepts independently. These critical thinking skills are transferable to other areas of life, equipping students with the ability to make informed decisions, solve complex problems, and approach challenges with confidence.

Real-World Application:

One of the key benefits of integrating 21st-century skills in math education is the increased emphasis on real-world application. Traditional math instruction often presents concepts in isolation, leading students to question the relevance of what they are learning. By incorporating real-world contexts and authentic problem-solving tasks, math classrooms become dynamic environments where students can see the practical applications of mathematics in their daily lives. This approach not only enhances students' understanding of mathematical concepts but also fosters a sense of relevance and motivation. Students are more likely to engage actively in learning when they can see how math connects to real-life situations, preparing them for the challenges they may encounter in their future careers and everyday decision-making.

Collaboration and Communication:

Integrating 21st-century skills in math education promotes collaboration and communication among students. Mathematics is often seen as an individual pursuit, with students working independently on problem sets. However, in the 21st century, collaboration is a vital skill for success in the workplace and beyond. By incorporating collaborative activities, group projects, and peer discussions, math classrooms provide opportunities for students to work together, share ideas, and learn from one another. Collaborative problem-solving tasks encourage students to listen to different perspectives, articulate their own reasoning, and engage in meaningful dialogue. This fosters not only a deeper understanding of mathematical concepts but also the development of essential communication and teamwork skills.

Integrating 21st-century skills in math education is paramount for preparing students for the challenges and opportunities of the digital age. By enhancing problem-solving abilities, cultivating critical thinking skills, emphasizing real-world application, and promoting collaboration and communication, math classrooms become dynamic and engaging spaces for students to develop the skills they need to succeed in an interconnected world. The integration of 21st-century skills not only strengthens students' mathematical proficiency but also equips them with transferable skills that are essential for lifelong learning, problem-solving, and success in the 21st century and beyond.

Pedagogical Approaches for Fostering 21st-Century Skills in Math

Project-Based Learning: Project-based learning (PBL) is a pedagogical approach that empowers students to apply their mathematical knowledge and skills to real-world problems. By engaging in hands-on projects, students actively participate in problem-solving, critical thinking, collaboration, and creative thinking. PBL provides a context for students to explore mathematical concepts in meaningful ways and develop a deeper understanding of their applications.

In a math classroom, a project-based learning approach could involve students working on a real-world challenge that requires them to apply mathematical principles. For example, students might design a city park, considering factors such as area, perimeter, and budget constraints. Through this project, they would engage in critical thinking as they analyse the dimensions of various park features, collaborate with their peers to brainstorm ideas, and creatively design a park that meets the specified requirements. The project could culminate in a presentation where students explain their design choices and mathematical reasoning.

Flipped Classroom Model:

The flipped classroom model involves the reversal of traditional instruction, where students learn new concepts independently at home through online resources, videos, or interactive simulations, and then engage in collaborative activities and discussions in the classroom. This approach allows students to take ownership of their learning, develop self-directed learning skills, and utilize technology for exploration and deeper understanding.

For instance, in a math classroom, students can watch pre-recorded instructional videos or interactive tutorials that introduce new mathematical concepts. They can then come to the classroom ready to engage in problem-solving activities, group discussions, and hands-on experiments that reinforce and apply the learned concepts. This model encourages students to actively participate in their learning process, seek clarification during classroom interactions, and develop critical thinking skills by working through challenging problems together.

Gamification in Math Education:

Gamification involves the integration of game elements and mechanics into educational contexts to enhance engagement, motivation, and learning outcomes. In math classrooms, gamification can be applied to foster problem-solving skills, digital literacy, and a growth mindset.

For example, a math teacher might introduce a gamified platform or app that presents math challenges in the form of quests, puzzles, or interactive simulations. Students earn points, badges, or rewards as they progress through levels or demonstrate mastery of mathematical skills. The game elements create a sense of achievement and friendly competition among students, motivating them to actively participate and persist in solving math problems. Gamification also allows for immediate feedback, adaptive learning experiences, and the integration of technology, promoting digital literacy and preparing students for the technology-driven world.

Blended Learning:

Blended learning combines face-to-face instruction with online resources and tools to create personalized and flexible learning experiences. In a math classroom, blended learning allows for a variety of instructional approaches, including online tutorials, interactive exercises, virtual simulations, and collaborative projects.

For instance, students can access online math platforms or applications that provide personalized learning pathways, adaptive assessments, and instant feedback. They can engage in self-paced activities to reinforce mathematical concepts, explore virtual manipulatives, and access additional resources for independent exploration. In the classroom, teachers can facilitate collaborative discussions, group projects, and problem-solving tasks that leverage both online and offline resources. Blended learning promotes student autonomy, encourages digital collaboration, and caters to individual learning needs and preferences.

These pedagogical approaches demonstrate how integrating 21st-century skills in math classrooms can be accomplished through project-based learning, the flipped classroom model, gamification, and blended learning. By adopting these approaches, educators can create dynamic and engaging learning environments that foster problem-solving, critical thinking, collaboration, creativity, and digital literacy. These examples provide a glimpse into how these approaches have been successfully implemented in real-life experiences of children, empowering them to develop essential 21st-century skills alongside their mathematical proficiency.

Technology Integration for 21st-Century Skills Development

In today's rapidly advancing world, the integration of technology in education has become crucial to prepare students for the challenges and opportunities of the 21st century. In the field of mathematics education, technology offers a wide range of tools and resources that can enhance learning experiences and foster the development of 21st-century skills. In this chapter, we will explore various aspects of technology integration in math classrooms, highlighting the significance of digital tools, data visualization and analysis, coding and robotics, and virtual reality and augmented reality.

Digital Tools and Apps for Math Education

Digital tools and applications provide innovative ways for students to engage with mathematical concepts and develop their problem-solving and critical thinking skills. These tools offer interactive simulations, virtual manipulatives, and adaptive learning platforms that can transform math learning experiences.

One example of a digital tool is GeoGebra, a dynamic mathematics software that allows students to explore geometric and algebraic concepts. With GeoGebra, students can manipulate shapes, graphs, and equations, providing them with a visual and interactive representation of mathematical relationships. By experimenting with different parameters and observing the effects on the graphs, students develop a deeper understanding of mathematical concepts and enhance their critical thinking skills.

Another digital tool that supports math learning is Desmos, an online graphing calculator. Desmos provides a platform for students to graph functions, explore transformations, and analyse data sets. Its user-friendly interface and interactive features enable students to actively participate in mathematical exploration and problem-solving. For example, students can use Desmos to graph real-world scenarios, such as population growth or financial trends, and analyse the data to make informed predictions and decisions.

Adaptive learning platforms, such as Khan Academy, provide personalized math instruction based on students' individual needs and progress. These platforms offer a vast array of math exercises and tutorials, allowing students to practice specific skills and receive immediate feedback. Adaptive learning platforms not only enhance students' mathematical proficiency but also foster self-directed learning, as students can work at their own pace and focus on areas where they need additional support.

By incorporating digital tools and apps into math classrooms, educators create engaging and interactive learning environments where students can actively explore, analyse, and apply mathematical concepts. These tools provide opportunities for students to develop problem-solving, critical thinking, and digital literacy skills, which are essential for success in the 21st century.

Data Visualization and Analysis

Data literacy and analysis have become increasingly important in our data-driven society. By utilizing data visualization tools, students can analyse and interpret mathematical data, thereby promoting critical thinking and problem-solving skills.

One powerful data visualization tool is Tableau Public, a free software that allows students to create interactive visualizations from raw data. Students can import data sets, manipulate variables, and design visual representations such as charts, graphs, and maps. For instance, students can use Tableau Public to analyse survey data on favorite sports in different countries and create a global heatmap to identify patterns and trends. By engaging in data visualization, students gain a deeper understanding of statistics and develop the ability to draw meaningful conclusions from data.

In addition to software tools, online platforms like Gapminder provide access to vast datasets and interactive visualizations that depict global trends and phenomena. By exploring these visualizations, students can develop data literacy skills and gain a global perspective on various mathematical concepts. For example, students can analyse population growth, income inequality, or carbon emissions over time, fostering critical thinking and awareness of real-world issues.

By incorporating data visualization tools into math classrooms, educators can empower students to become data-literate individuals who can effectively analyse and interpret data. Through these activities, students develop critical thinking skills, make evidence-based decisions, and become informed consumers of information in the digital age.

Coding and Robotics in Math

The integration of coding and robotics in math classrooms provides unique opportunities for students to develop computational thinking, algorithmic problem-solving, and logical reasoning skills. By engaging in coding activities, students learn to apply mathematical concepts in practical and meaningful ways.

Coding platforms like Scratch and Blockly enable students to create animations, games, and interactive stories by programming characters and objects. Through these coding projects, students develop a deep understanding of mathematical concepts such as variables, loops, and conditional statements. For instance, students can use variables to control the speed and direction of a character's movement or implement mathematical formulas to calculate scores or distances in a game. By coding, students become active participants in their learning, applying mathematical knowledge to solve problems and express their creativity.

Robotics provides another avenue for integrating math and computational thinking. Students can program robots to perform specific tasks that involve mathematical concepts. For example, students can program a robot to navigate a maze, calculate distances using sensors, or solve mathematical puzzles. Through these activities, students develop logical reasoning skills, learn to debug and iterate their code, and understand the practical applications of mathematics in real-world scenarios.

The integration of coding and robotics in math classrooms not only enhances students' mathematical understanding but also nurtures their problem-solving, critical thinking, and collaboration skills. By engaging in hands-on coding and robotics activities, students develop the skills necessary to thrive in a technology-driven society.

Virtual Reality and Augmented Reality

Virtual reality (VR) and augmented reality (AR) technologies have the potential to revolutionize math education by creating immersive and interactive learning experiences. These technologies allow students to visualize complex mathematical concepts, explore three-dimensional environments, and make connections between abstract ideas and real-world applications.

With VR, students can explore virtual worlds that represent mathematical concepts in an engaging and interactive manner. For example, students can enter a virtual geometry environment where they can manipulate shapes, explore spatial relationships, and discover geometric properties. By immersing themselves in a three-dimensional space, students develop a deeper understanding of geometry concepts and enhance their spatial reasoning skills.

AR, on the other hand, overlays digital content onto the real world, providing students with interactive and context-rich experiences. For instance, students can use AR apps to measure angles in their physical surroundings, visualize three-dimensional graphs, or solve math puzzles by interacting with virtual objects in their immediate environment. AR brings abstract mathematical concepts to life, making them more tangible and accessible to students.

Looking to the future, the potential of VR and AR in math education is vast. As these technologies continue to evolve, students could engage in collaborative VR environments where they solve math problems together, simulate real-world mathematical scenarios, or explore mathematical concepts in AR-enhanced textbooks or learning materials. These technologies promote spatial reasoning, problem-solving, creativity, and digital literacy, preparing students for the digital landscape they will encounter in the 21st century.

The integration of technology in math education offers numerous opportunities for fostering 21st-century skills. By incorporating digital tools and apps, data visualization and analysis, coding and robotics, and virtual reality and augmented reality, educators can create dynamic and engaging learning environments that empower students to develop problem-solving, critical thinking, collaboration, and digital literacy skills. These technological advancements not only enhance students' mathematical proficiency but also prepare them for the challenges and opportunities of the rapidly evolving digital age. By embracing these pedagogical approaches, educators can ensure that math classrooms reflect the use of advanced technology and devices, providing students with the skills they need to succeed in the 21st century and beyond.

Integrating 21st-century skills in math classrooms is imperative for preparing students for the digital age. By embracing advanced technology and employing pedagogical approaches that foster critical thinking, problem-solving, collaboration, creativity, and digital literacy, educators can create dynamic and impactful learning experiences. I hope, this canto has provided practical guidance and examples to inspire educators to embrace these strategies, equipping students with the necessary skills to thrive in a rapidly evolving world.

CANTO EIGHT

Celestial Explorations: Scientific Brilliance Unveiled

In the rapidly evolving world of the 21st century, education must adapt to equip students with the skills they need to thrive in a complex and interconnected society. This chapter focuses on the integration of 21st-century skills in science classrooms, highlighting the significance of critical thinking, problem-solving, collaboration, communication, creativity, and digital literacy. We will explore various pedagogical approaches and strategies that educators can employ to effectively foster these skills in science education. By providing practical guidance and examples, this chapter aims to support educators in creating engaging and innovative learning environments that prepare students for the challenges of the modern world.

Understanding 21st-Century Skills in Science Education

Defining 21st-Century Skills in Science

Before delving into the pedagogical approaches, it is essential to establish a clear understanding of the 21st-century skills relevant to science education. Critical thinking, problem-solving, collaboration, communication, creativity, and digital literacy are the key skills that students need to develop. In the context of science, critical thinking involves evaluating evidence, analysing data, and making informed decisions. Problem-solving entails applying scientific knowledge and methodologies to address real-world issues. Collaboration fosters teamwork and the ability to work effectively with others to solve complex problems. Communication involves effectively conveying scientific ideas and concepts to various audiences. Creativity encourages students to think outside the box, generate innovative ideas, and approach scientific challenges with an open mind. Digital literacy encompasses the ability to access, evaluate, and use digital information effectively.

The Relevance of 21st-Century Skills in Science Education

The integration of 21st-century skills in science classrooms is vital for several reasons. First, science is an inherently inquiry-based discipline that requires critical thinking and problem-solving. By fostering these skills, students develop a deeper understanding of scientific concepts and the ability to apply them in real-world contexts. Second, collaboration is essential in science as scientists often work in teams to conduct research and solve complex problems. By engaging students in collaborative activities, they learn to value diverse perspectives, communicate effectively, and contribute meaningfully to a shared goal. Third, effective communication is crucial in science as scientists need to articulate their findings, write research papers, and present their work to peers and the broader scientific community. By honing their communication skills, students become effective science communicators who can disseminate knowledge and engage in scientific discourse. Fourth, fostering creativity in science classrooms encourages students to think critically, explore new ideas, and develop innovative solutions to scientific challenges. Lastly, digital literacy is paramount in the digital age as students must navigate vast amounts of scientific information, utilize technology tools, and analyse data effectively.

Pedagogical Approaches for Fostering 21st-Century Skills in Science

Inquiry-Based Learning

Inquiry-based learning is a pedagogical approach that encourages students to investigate scientific phenomena, ask questions, and construct their own understanding of scientific concepts. By engaging in hands-on experiments, conducting research, and analysing data, students develop critical thinking, problem-solving, collaboration, and communication skills. Educators can implement inquiry-based learning by designing open-ended investigations and providing guidance and support as students explore scientific phenomena. For example, students can investigate the effect of different variables on plant growth, design experiments to test the properties of materials, or explore the impact of human activities on the environment. Through inquiry-based learning, students actively participate in the scientific process, fostering a deeper understanding of scientific concepts and the development of 21st-century skills.

Project-Based Learning

Project-based learning is another effective approach for fostering 21st-century skills in science classrooms. By engaging in long-term, student-driven projects, students tackle real-world problems, collaborate with peers, and apply scientific knowledge to propose solutions. For example, students can design a sustainable energy project, investigate the impact of pollution on local ecosystems, or develop a plan to address a community health issue. Throughout the project, students develop critical thinking, problem-solving, collaboration, communication, and creativity skills. Educators can facilitate project-based learning by providing guidance, scaffolding, and resources, while allowing students to take ownership of their learning and make meaningful connections between science and the real world.

STEM Integration

Integrating science, technology, engineering, and mathematics (STEM) provides a holistic approach to fostering 21st-century skills. By connecting these disciplines, students engage in interdisciplinary problem-solving and develop a comprehensive skill set. For example, students can design and build models to solve engineering challenges, utilize technology tools to collect and analyse scientific data, and apply mathematical concepts to solve scientific problems. STEM integration promotes critical thinking, problem-solving, collaboration, communication, and digital literacy. Educators can facilitate STEM integration by designing projects and activities that bridge multiple disciplines and encourage students to apply their knowledge and skills in an integrated manner.

Use of Technology and Digital Tools

Incorporating technology and digital tools in science classrooms enhances students' digital literacy and supports the development of other 21st-century skills. Educators can leverage various tools such as simulations, virtual labs, data analysis software, and online collaboration platforms. For example, students can use simulation software to explore the impact of variables on a virtual ecosystem, collect and analyse real-time data using sensors, or collaborate with peers on a shared online platform to solve scientific problems. By using technology, students develop digital literacy skills, enhance their problem-solving abilities, and gain exposure to cutting-edge scientific tools and methodologies.

Step 1: Identify the skills

The first step is to identify the specific 21st-century skills that are relevant to the science curriculum and the desired learning outcomes. By conducting a skills inventory or assessment, educators can determine the existing strengths and areas for improvement in students' skills. This can be achieved through self-reflection, surveys, or direct observation. Additionally, teachers can provide explicit instruction and guidance on the specific skills they expect students to develop throughout the science curriculum.

For example, in a high school biology class, educators may identify critical thinking, collaboration, and digital literacy as key skills. They can emphasize the importance of analysing scientific information, working effectively in teams, and utilizing technology tools for research and data analysis.

Step 2: Align the curriculum

The second step involves aligning the existing science curriculum with the identified 21st-century skills. Educators should analyse the curriculum and identify opportunities for integrating these skills into the instructional design. By selecting concepts or topics that lend themselves to collaborative, inquiry-based, or project-based learning, educators can create meaningful connections between the content and the development of 21st-century skills.

For instance, in a middle school physics class, educators may align the curriculum with collaboration and problem-solving skills by designing a unit on designing and building structures. Students could work in teams to construct bridges or towers using principles of physics, requiring them to collaborate, think critically, and apply their scientific knowledge to real-world challenges.

Step 3: Design learning experiences

The next step is to design learning experiences that promote the development of 21st-century skills. Educators should consider incorporating inquiry-based investigations, project-based learning activities, STEM integration, and technology tools to create engaging and interactive experiences for students. These experiences should provide opportunities for critical thinking, collaboration, creativity, and communication.

In a high school chemistry class, for example, educators can design an inquiry-based experiment where students investigate the factors affecting the rate of a chemical reaction. By working in groups, students can collaborate to design and conduct experiments, analyse data, and draw conclusions. This experience not only develops their critical thinking and collaboration skills but also enhances their understanding of chemical principles.

Step 4: Provide guidance and support

To facilitate the learning process and support students' development of 21st-century skills, educators need to provide guidance, scaffolding, and resources. Differentiated instruction can be implemented to meet the diverse needs of students in developing these skills. Educators can offer modelling and scaffolding techniques, such as think-alouds, to demonstrate critical thinking processes. They can provide templates or graphic organizers to support collaborative work, and offer technological tools and resources to enhance students' digital literacy and information literacy skills.

For instance, in an elementary school earth science class, educators can guide students in using online resources and interactive simulations to explore natural disasters. By modeling effective search strategies, providing graphic organizers to structure their research, and facilitating discussions, educators can support students in developing their critical thinking, collaboration, and digital literacy skills.

Step 5: Assess and provide feedback

Assessment plays a vital role in measuring students' progress in developing 21st-century skills. Educators should utilize a variety of formative and summative assessment methods to assess these skills effectively. Rubrics or checklists can be used to evaluate critical thinking, collaboration, and communication skills demonstrated in class discussions, group projects, or scientific presentations. It is essential to provide timely and constructive feedback to guide students' growth, highlighting areas for improvement and suggesting strategies for further development.

For example, in a middle school environmental science class, educators can assess students' collaboration skills during a group project on designing sustainable solutions for a local environmental issue. By observing their teamwork, communication, and problem-solving abilities, educators can provide feedback that focuses on specific areas of growth and suggests strategies for enhancing their collaborative skills.

Step 6: Reflect and refine

The final step in the process is to encourage reflection and refinement. Educators should provide regular opportunities for students to reflect on their growth in 21st-century skills and their transferability to real-life situations. By engaging in metacognition, students become aware of their own learning processes and can better understand the value of these skills beyond the science classroom.

Moreover, educators should also reflect on their instructional practices and seek feedback from students, colleagues, and other stakeholders. Professional learning communities or workshops can be utilized to share experiences, exchange ideas, and refine instructional strategies for integrating 21st-century skills effectively in science classrooms. By continuously reflecting and refining their approaches, educators can ensure that they are providing the best possible learning experiences for their students.

As the world undergoes rapid technological advancements, it is imperative for science classrooms to evolve and embrace these changes. By integrating technology into science education, classrooms can provide students with authentic and engaging learning experiences that reflect the realities of the 21st century.

The Process Skills

Let's see how the process skills connected with science learning and the 21st century skills be integrated to equip learners with scientific temperament in science classrooms.

Observation:

Observation is the process of using the senses or technology to gather information about the world. In a science classroom, technology can enhance students' observational skills by providing access to detailed and real-time data. For instance, students can use digital microscopes to observe the structure of cells or use temperature sensors to monitor changes in heat levels during chemical reactions. By utilizing technology tools, students can make more accurate and precise observations, leading to a deeper understanding of scientific phenomena.

Example: In a biology class, students can use a digital microscope connected to a computer or tablet to examine different types of plant cells. They can observe the variations in cell structure, identify organelles, and compare the differences between plant cells under different conditions. This technology enables students to view cells at a higher magnification and make more detailed observations than with traditional microscopes.

Comparison:

Comparison involves examining the similarities and differences between objects, phenomena, or data. Technology provides students with access to vast amounts of information and resources, enabling them to make comparisons easily and efficiently. Online databases, interactive visuals, and digital simulations allow students to compare variables, graphs, or scientific models to identify patterns and draw meaningful conclusions.

Example: In a physics class, students can use a simulation software to compare the motion of objects under different conditions, such as varying mass or friction. By adjusting the variables and analysing the resulting data, students can compare the effects of different factors on the motion of objects, identifying relationships and patterns. This technology allows for quick and repeated comparisons, enhancing students' understanding of the underlying principles of motion.

Classification:

Classification involves categorizing objects, organisms, or data based on their shared characteristics. Technology can support students in the process of classification by providing access to databases, digital field guides, or image recognition tools. These resources help students identify and classify species, minerals, or other scientific entities accurately and efficiently.

Example: In a biology class, students can use an online database or app to identify and classify different species of birds. By uploading an image of a bird, the image recognition technology can provide information about the species, its habitat, and its unique characteristics. This technology assists students in accurately classifying the birds they observe in the field and expands their knowledge of different bird species.

Analysis:

Analysis involves examining data, patterns, or relationships to derive meaningful insights. Technology offers powerful tools for data analysis, such as statistical software, spreadsheet applications, and graphing tools. These tools enable students to analyse and interpret scientific data more effectively, helping them draw conclusions and make evidence-based claims.

Example: In a chemistry class, students can use a spreadsheet application to analyse the relationship between temperature and the rate of a chemical reaction. By plotting the data points on a graph and using the software's analysis functions, students can identify trends, calculate reaction rates, and draw conclusions about the relationship between temperature and reaction rate. This technology supports students in analysing complex data sets and developing a deeper understanding of chemical reactions.

Synthesis:

Synthesis involves integrating information, ideas, or concepts to create a cohesive understanding. Technology provides platforms for students to collaborate, share ideas, and synthesize their findings. Online collaboration tools, discussion forums, or shared document platforms allow students to work together, exchange perspectives, and integrate diverse insights to develop comprehensive understandings of scientific concepts.

Example: In an environmental science class, students can collaborate on a shared document or online platform to synthesize their research on the impacts of human activities on local ecosystems. Each student can contribute their findings, analysis, and proposed solutions, allowing for a comprehensive synthesis of information. This technology promotes collaboration, critical thinking, and the integration of multiple perspectives to address complex environmental issues.

Application:

Application involves using scientific knowledge and skills to solve real-world problems or make predictions. Technology provides platforms for students to apply their scientific understanding in virtual experiments or simulations. Virtual simulations, modelling software, or interactive apps allow students to engage in realistic scenarios where they can apply their scientific knowledge and skills.

Example: In a geology class, students can use a virtual simulation to study the effects of erosion on landforms. By adjusting variables such as rainfall intensity or slope steepness, students can observe the resulting changes in the landscape, such as the formation of canyons or the deposition of sediment. This technology enables students to apply their understanding of erosion processes in a virtual environment, fostering critical thinking and problem-solving skills.

In a science classroom where 21st-century skills and process skills are integrated, the role of the science teacher is vital. The teacher serves as a facilitator and guide, helping students navigate the use of technology effectively to develop and enhance process skills. They provide support by introducing technology tools, demonstrating their use, and guiding students in applying these tools to specific scientific tasks. The teacher also encourages critical thinking, inquiry, and reflection, promoting the development of process skills in students. The science teacher creates a supportive and inclusive classroom environment that fosters collaboration, creativity, and independent thinking. They design authentic learning experiences that require students to apply their process skills in real-world contexts, challenging them to solve problems, conduct investigations, and make connections to everyday life. By integrating 21st-century skills and process skills, the science teacher empowers students to become active learners, critical thinkers, and scientifically literate individuals prepared for the challenges of the rapidly evolving world.

Integrating 21st-century skills in science classrooms is crucial for preparing students to thrive in the modern world. By incorporating pedagogical approaches such as inquiry-based learning, project-based learning, STEM integration, and the use of technology and digital tools, educators can create engaging and innovative learning environments that foster critical thinking, problem-solving, collaboration, communication, creativity, and digital literacy. Following a step-by-step process that involves identifying skills, aligning the curriculum, designing learning experiences, providing guidance and support, assessing and providing feedback, and reflecting and refining, educators can effectively integrate 21st-century skills in science education. By equipping students with these essential skills, we empower them to become scientifically literate, adaptable, and active participants in our ever-changing world.

CANTO NINE

Horizon's Whispers: Shaping the Future's Canvas

The 21st century has brought forth a wave of technological advancements, transforming the way we live, work, and communicate. In this era of rapid change, the development of 21st-century skills has become imperative to prepare individuals for the challenges and opportunities that lie ahead. These skills, encompassing critical thinking, collaboration, creativity, communication, digital literacy, and adaptability, are not only essential for success in education but also for thriving in the workplace and contributing to a dynamic society. This chapter delves into the implications of 21st-century skills for the future, exploring their influence on education, the workplace, and society as a whole.

The Changing Landscape of Education:

Over the past decade, the integration of 21st-century skills in education has gained significant momentum, reshaping the way teaching and learning take place in classrooms. Traditional approaches that relied heavily on rote memorization and passive learning are gradually being replaced by student-centered, inquiry-based, and technology-driven models. This shift is driven by the recognition that 21st-century skills are essential for students to thrive in an increasingly interconnected and complex world.

One example of the transformation in education is the widespread adoption of project-based learning (PBL). PBL provides students with opportunities to apply their knowledge and skills to real-world problems and challenges. By engaging in hands-on projects, students develop critical thinking, collaboration, communication, and problem-solving skills. For instance, in a science class, students might work in groups to design and conduct experiments to investigate the impact of pollution on local ecosystems. Through this process, they not only deepen their understanding of scientific concepts but also develop essential 21st-century skills by working collaboratively, analysing data, and communicating their findings.

Personalized learning is another approach that has gained traction in recent years. With advancements in technology, educators can now leverage digital tools and adaptive learning platforms to tailor instruction to individual students' needs, interests, and learning styles. For example, a math teacher might use an online platform that provides personalized practice exercises based on each student's performance and learning pace. This approach allows students to progress at their own speed, receive immediate feedback, and engage in self-directed learning, promoting critical thinking, self-motivation, and digital literacy.

The flipped classroom model has also become increasingly popular. In this model, students learn new content at home through video lectures or online resources, and class time is then dedicated to collaborative activities, discussions, and problem-solving. This approach allows for more interactive and student-centered learning experiences, where students actively apply their knowledge and skills. For instance, in a flipped science classroom, students might watch a video lecture about the scientific method at home and then come to class ready to conduct experiments, analyse data, and discuss their findings. This fosters critical thinking, collaboration, and communication skills, as students engage in scientific inquiry and reasoning.

Looking ahead, it is expected that the integration of 21st-century skills in education will continue to evolve and become even more embedded in educational systems. In the next 20 years or so, we can envision a future where classrooms are equipped with advanced technologies such as virtual reality (VR) and augmented reality (AR), enabling immersive and interactive learning experiences. For instance, students could use VR to explore the human body in a biology class or travel back in time to experience historical events. Such experiences would not only enhance students' understanding of content but also foster creativity, adaptability, and global citizenship as they engage with diverse perspectives and cultures.

Additionally, advancements in artificial intelligence (AI) and machine learning hold great potential for personalized and adaptive learning. AI-powered tools could provide instant feedback, analyse individual learning patterns, and offer tailored recommendations for each student. For example, an AI tutor could adapt its instruction based on a student's strengths and weaknesses, providing targeted resources and challenges to support their learning journey. This personalized approach would empower students to take ownership of their learning, develop self-directed learning skills, and cultivate resilience and adaptability.

In this envisioned future, education would increasingly emphasize the development of higher-order thinking skills, creativity, and problem-solving abilities. The focus would shift from memorization of facts to the application of knowledge in complex and authentic contexts. Students would be encouraged to think critically, analyse information, solve problems creatively, and communicate effectively across different mediums. This holistic approach to education would nurture lifelong learners who are adaptable, innovative, and equipped to navigate the challenges and opportunities of the 21st century.

Integration of 21st-century skills in education has already begun to reshape the learning landscape, with traditional approaches giving way to student-centered, inquiry-based, and technology-driven models. Examples such as project-based learning, personalized learning, and the flipped classroom demonstrate the transformative impact of these approaches on promoting critical thinking, collaboration, and digital literacy. Looking ahead, the future of education will likely witness further advancements in technology, such as VR, AR, and AI, which will enhance personalized learning and foster the development of 21st-century skills. By equipping students with the skills and mindsets necessary to thrive in an ever-evolving world, education will continue to prepare individuals for lifelong learning, innovation, and global citizenship.

The Evolving Workplace:

The rapid advancements in automation and artificial intelligence are reshaping the workforce, creating a demand for a new set of skills and competencies. As technology continues to automate routine tasks, the jobs of the future will require a combination of technical expertise and 21st-century skills that are uniquely human.

The ability to think critically, solve complex problems, collaborate across disciplines, and adapt to change will be crucial in navigating the evolving workplace.

One example of the changing landscape is the field of healthcare. With the development of AI-powered diagnostic tools and robotic assistance in surgeries, healthcare professionals are increasingly expected to possess a strong foundation in technical knowledge. However, alongside this expertise, they will also need to demonstrate critical thinking skills to analyse patient data, make informed decisions, and provide personalized care. Effective communication and collaboration will be essential for interdisciplinary teamwork, where doctors, nurses, technicians, and AI systems work together to deliver the best outcomes for patients. Additionally, adaptability will be vital as healthcare professionals will need to keep pace with rapidly evolving technologies and treatment methods.

In the realm of entrepreneurship and innovation, the future workplace will require individuals who possess not only technical skills but also a strong entrepreneurial mindset. Entrepreneurs will need to identify market opportunities, develop innovative solutions, and navigate a dynamic business environment. Creativity will play a central role in generating new ideas, while critical thinking will be essential for evaluating risks and making strategic decisions. Collaboration and networking will be crucial for building partnerships and accessing diverse resources. Moreover, adaptability will enable entrepreneurs to pivot and adjust their strategies as market conditions change.

Another significant impact of 21st-century skills on the workplace is the increasing reliance on digital tools and platforms. As technology continues to advance, the ability to effectively navigate and leverage digital resources will be essential across industries.

For example, data analysis and visualization tools will enable professionals to derive insights from vast amounts of information, leading to informed decision-making. Digital collaboration platforms will facilitate remote teamwork, enabling individuals from different locations to collaborate seamlessly. Virtual reality and augmented reality technologies may revolutionize training and simulations, providing immersive learning experiences that enhance skill development.

Looking ahead to the future, it is anticipated that the demand for 21st-century skills will only increase. As automation and AI continue to advance, jobs that require repetitive tasks are likely to be automated, leading to a shift in the types of jobs available. The jobs of the future will place a greater emphasis on skills that are uniquely human, such as creativity, critical thinking, emotional intelligence, and adaptability. This means that individuals who possess these skills will be well-positioned to thrive in the evolving workplace.

The concept of "workforce readiness" is being redefined in light of 21st-century skills. Traditionally, workforce readiness focused primarily on technical skills and knowledge specific to a particular job. However, as the workplace evolves, employers are recognizing the need for individuals who can adapt to changing circumstances, think critically, communicate effectively, and collaborate with diverse teams. This shift in mindset has led to a greater emphasis on the development of transferable skills that can be applied across various roles and industries.

Integration of 21st-century skills in education is preparing individuals for the evolving workplace. The combination of technical expertise and skills such as critical thinking, collaboration, adaptability, and digital literacy will be essential for success in the jobs of the future. Examples from healthcare, entrepreneurship, and the increasing reliance on digital tools demonstrate how these skills are becoming increasingly important. As automation and AI continue to advance, the demand for 21st-century skills will continue to grow, reshaping the concept of workforce readiness and redefining the skills needed to thrive in the workplace of the future.

Embracing Technology:

Technology plays a vital role in fostering 21st-century skills by providing tools and platforms that facilitate collaboration, critical thinking, and creativity. Educators can leverage technology to create interactive learning experiences, engage students in real-world problem-solving, and enhance digital literacy. For example, virtual reality simulations can immerse students in authentic scientific explorations, while online collaboration tools can enable global connections and knowledge sharing. As technology continues to advance, it will become even more integral to the learning process, offering new possibilities for skill development.

Providing Professional Development

Educators require ongoing professional development opportunities to effectively integrate 21st-century skills into their teaching practices. Professional development programs should focus on equipping teachers with the necessary knowledge and strategies to foster critical thinking, collaboration, communication, and creativity in the classroom. For instance, workshops and training sessions can introduce educators to innovative teaching methodologies, provide hands-on experience with relevant technologies, and encourage collaboration among teachers to share best practices. By investing in professional development, educators can effectively guide students in acquiring and applying 21st-century skills.

Implementing Authentic Assessments:

Traditional assessments often fail to capture the full range of 21st-century skills. Educators should explore alternative assessment methods that reflect real-world applications of these skills. Performance-based assessments, portfolios, and project-based evaluations allow students to demonstrate their abilities in problem-solving, critical thinking, and communication. For instance, instead of a traditional written exam, students can be tasked with designing and presenting a solution to a complex problem, showcasing their ability to apply knowledge and collaborate effectively.

Fostering Collaboration:

Collaboration is a fundamental 21st-century skill that prepares students for the collaborative nature of the modern workplace. Educators should create opportunities for students to work together on projects, engage in discussions, and solve problems collectively. Collaborative activities promote effective communication, teamwork, and the sharing of diverse perspectives. For example, group projects that require students to research, analyse data, and present findings not only develop their collaboration skills but also enhance their critical thinking and communication abilities.

Bridging Education and Industry:

To ensure the relevance and applicability of 21st-century skills, it is crucial to establish partnerships between educational institutions and industry. This collaboration can take various forms, such as internships, mentorship programs, and guest lectures by professionals from different fields. By connecting students with real-world contexts, they gain a deeper understanding of how their skills are valued and utilized in professional settings. Additionally, industry professionals can provide insights into emerging trends and skill requirements, helping educators align their curricula with the needs of the future workforce.

Integration of 21st-century skills into education is expected to continue evolving. After 20 years, it is anticipated that educational systems will have fully embraced the importance of these skills, resulting in widespread implementation and integration across all subject areas. The use of technology will be even more advanced, with virtual reality, artificial intelligence, and personalized learning platforms becoming commonplace. Professional development programs will continue to support educators in adapting their teaching practices to meet the changing demands of education. Assessments will become more authentic, focusing on students' abilities to apply knowledge and skills in real-world contexts. Collaboration will remain a cornerstone of education, with students engaging in global partnerships and interdisciplinary projects. Lastly, the strong collaboration between education and industry will ensure that students are well-prepared for the demands of the workforce, and lifelong learning will be embraced as individuals continuously adapt to an ever-changing world.

The implications of 21st-century skills for the future are far-reaching. These skills not only empower individuals to navigate the complexities of the digital age but also have a transformative impact on education, the workplace, and society as a whole. As we strive to prepare the next generation for an uncertain future, the integration of 21st-century skills becomes a fundamental component of educational systems worldwide. By equipping learners with the tools and mindsets necessary to thrive in an ever-changing world, we can collectively shape a future that embraces innovation, collaboration, and the pursuit of knowledge.

Societal Impact:

The development of 21st-century skills has the potential to reshape society in significant ways. As individuals become proficient in critical thinking, collaboration, and communication, they are better equipped to address complex societal challenges and drive positive change. Here are some examples of the societal impact of 21st-century skills:

Social Innovation:

21st-century skills enable individuals to think creatively and develop innovative solutions to social problems. By fostering creativity, problem-solving, and entrepreneurial thinking, individuals can address issues such as poverty, inequality, and environmental sustainability. For instance, imagine a future where students equipped with 21st-century skills collaborate to develop sustainable technologies, design social enterprises, or create initiatives to promote equality and inclusivity.

Civic Engagement:

Active participation in civic life is crucial for a healthy democracy. 21st-century skills empower individuals to engage in informed decision-making, advocate for their rights, and contribute to the well-being of their communities. With strong critical thinking, communication, and collaboration skills, individuals can engage in meaningful dialogue, work together to address social issues, and promote social justice. In the future, we can envision citizens who actively participate in community projects, engage in policy discussions, and use their skills to create positive change.

Sustainable Development:

As the world grapples with environmental challenges, 21st-century skills can play a pivotal role in promoting sustainable development. Individuals who possess critical thinking, problem-solving, and systems thinking skills can contribute to finding innovative solutions to issues such as climate change, resource depletion, and pollution. For example, students equipped with these skills may collaborate to develop alternative energy sources, design eco-friendly technologies, or create sustainable agricultural practices.

Technology Integration:

Technology will continue to advance rapidly, and its integration in education will be even more prominent after 20 years. Emerging technologies such as artificial intelligence, virtual reality, and augmented reality will play an integral role in facilitating personalized and immersive learning experiences. Students will have access to interactive simulations, virtual laboratories, and AI-powered adaptive learning platforms, enhancing their critical thinking, problem-solving, and digital literacy skills.

Lifelong Learning:

The rapid pace of technological advancements and societal changes requires individuals to embrace lifelong learning. Educators and policymakers must promote a culture of continuous learning, encouraging individuals to develop a growth mindset and adapt to new challenges. Future educational systems may incorporate flexible learning pathways, micro-credentials, and online platforms that provide opportunities for individuals to upskill and reskill throughout their lives.

Collaboration with Industry:

Collaboration between education and industry will be essential to ensure that education aligns with the demands of the workforce. Partnerships with industry professionals, internships, and work-based learning experiences will bridge the gap between classroom learning and real-world application. This collaboration will enable students to gain practical skills, industry knowledge, and professional networks, preparing them for successful transitions into the workforce.

Ethical Considerations:

As technology continues to advance, ethical considerations surrounding its integration into education will become more prominent. Educators and policymakers must navigate issues such as data privacy, digital citizenship, and equity in access to technology. A balanced approach that addresses these ethical concerns while harnessing the potential of technology will be crucial to ensure that 21st-century skills are developed in an ethical and inclusive manner.

The societal impact of 21st-century skills will be transformative. As individuals equipped with these skills contribute to social innovation, civic engagement, and sustainable development, societies will become more resilient, inclusive, and prosperous. The future of education will embrace technology, lifelong learning, collaboration with industry, and ethical considerations to empower individuals with the skills needed to thrive in a rapidly evolving world.

CANTO TEN

Epilogue: Beyond the Stars, A Journey Unfolds

In a world where imagination dances with boundless possibilities, where the symphony of progress and innovation orchestrates a mesmerizing melody, we dare to dream beyond the realms of convention. Like a poetic crescendo that rises to meet the stars, we weave a tapestry of wild and audacious policy decisions, envisioning an education landscape that harmonizes with the pulse of the future. With each verse, we paint a portrait of a world transformed, where the rhythms of 21st-century skills intertwine with the melody of possibility, and the dance of academia embraces the ever-evolving tempo of a rapidly changing world. As we embark on this rhythmic journey, let us immerse ourselves in the cadence of visionary policy decisions, daring to dream higher and reach for the celestial realms of tomorrow.

As we conclude this book, let us be on board on a visionary journey into the future, where the development of 21st-century skills in learners has reshaped our world in profound and extraordinary ways. This chapter dares to dream and imagine a future where the integration of technology, the evolution of education, and the fostering of 21st-century skills have created a society that thrives on innovation, collaboration, and social progress. Through this exploration, we aim to inspire and ignite a sense of possibility, urging educators, policymakers, and stakeholders to continue pushing the boundaries of education and cultivate a future that transcends our current expectations.

Technological Marvels:

After 20 years, technology will have advanced beyond our wildest imagination, revolutionizing the learning experience and transforming the role of classrooms. With holographic displays, mind-machine interfaces, and immersive virtual reality experiences, students will be immersed in mind-bending simulations and experiments that push the boundaries of scientific understanding. The fusion of physical and digital realities will open up endless possibilities for learning and exploration. Learners will harness the power of advanced technologies to solve complex problems, ignite their creativity, and unlock their full potential.

Transformative Workforce:

The future workforce will be characterized by a dynamic ecosystem of creativity, entrepreneurship, and interdisciplinary collaboration. Emerging fields such as biotechnology, space exploration, and sustainable energy will be at the forefront, driven by a new generation of innovative thinkers and problem solvers.

Workplaces will embrace a culture of continuous learning and experimentation, allowing individuals to seamlessly transition between projects and domains as they pursue their passions and contribute to society's advancement. The development of 21st-century skills will be integral to this workforce, empowering individuals to navigate the ever-changing landscape of technology and drive innovation.

Global Unity and Cultural Fusion:

In the world after 20 years, 21st-century skills will have fostered a deep sense of global unity and cultural fusion. Virtual exchange programs will enable students from different corners of the globe to collaborate effortlessly, breaking down barriers of language and distance. Cultural diversity will be celebrated and integrated into educational experiences, cultivating a rich array of perspectives and fostering empathy, understanding, and interconnectedness. Through collaborative projects, learners will develop a global mindset and acquire the skills needed to address complex global challenges.

Personalized Learning 2.0:

The concept of personalized learning will have evolved to new heights, leveraging advanced artificial intelligence (AI) systems that intuitively adapt to each learner's unique needs and interests. Personalized learning assistants will guide students on their learning journey, providing tailored resources, real-time feedback, and adaptive assessments. Learners will have access to a vast array of learning pathways, expert mentors, and immersive learning experiences that cater to their individual strengths and aspirations. The integration of technology and data analytics will allow educators to gain deep insights into learners' progress, enabling personalized interventions and support.

Collaborative Innovation:

The power of collaboration will be amplified as individuals work together across disciplines, borders, and generations to address the grand challenges of humanity. Global collaborative platforms will empower students to co-create solutions for issues like climate change, poverty eradication, and equitable access to education. The focus will be on fostering a collective intelligence that transcends individual limitations, resulting in groundbreaking innovations and breakthroughs. The ability to collaborate effectively and leverage diverse perspectives will become a hallmark of 21st-century skills, driving societal progress and positive change.

Ethical Tech and Sustainable Futures:

As the integration of technology becomes more pervasive, ethical considerations and sustainability will play a central role in shaping the future. Students will be educated on the ethical implications of emerging technologies such as artificial intelligence, genetic engineering, and nanotechnology. They will develop a strong sense of digital citizenship and environmental stewardship, using their skills to design solutions that prioritize the well-being of both humanity and the planet. The integration of ethics and sustainability into the fabric of education will ensure responsible and thoughtful technology use, fostering a future that balances progress with societal and environmental considerations.

Lifelong Learning for Flourishing Lives:

In the world after 20 years, the pursuit of knowledge and personal growth will be lifelong endeavours. Education will extend far beyond the confines of formal institutions, encompassing a wide range of learning opportunities, from online courses and immersive experiences to community-driven initiatives. Individuals will embrace a growth mindset, continually seeking to acquire new skills, adapt to emerging technologies, and explore their passions. The idea of education as a transformative and lifelong journey will be deeply ingrained in society, enabling individuals to lead purposeful, fulfilling lives in a rapidly evolving world.

As we look ahead to a future where the development of 21st-century skills has ushered in a profound transformation, it is important to recognize that these visions are both aspirational and speculative. The realization of such a future will require boldness, collaboration, and ongoing dedication to innovation in education. By nurturing 21st-century skills, embracing technology, and fostering a culture of lifelong learning, we can create a world that exceeds our current expectations—a world where creativity, collaboration, and empathy are at the core of societal progress. Let us dare to dream, act with determination, and together shape a future that empowers learners to thrive and make a positive difference in the world.

Let's take our imagination to even greater heights and explore wild and futuristic policy decisions that could reshape education and society:

Mind-Uploading and Consciousness Transfer:

Academicians and policymakers could invest in research and development to explore the possibility of mind-uploading and consciousness transfer. This radical policy decision would enable individuals to transfer their consciousness into digital or synthetic forms, allowing for lifelong learning and personal growth beyond the limitations of the physical body. Learners could continue to develop their 21st-century skills in virtual realms, engaging in intellectual pursuits and collaborating with other digitally uploaded minds.

Neurotechnology's for Enhanced Learning:

Policymakers and academicians could embrace advancements in neurotechnologies to enhance the learning process. By using brain-computer interfaces (BCIs) or neurofeedback systems, learners could access information directly from their brains, improving memory, cognition, and information processing. This policy decision would revolutionize the way students acquire knowledge and develop 21st-century skills, opening up new frontiers of human potential.

Multi-Dimensional Learning Spaces:

Academicians and policymakers could revolutionize traditional classroom environments by creating multi-dimensional learning spaces. Through the use of holographic projections, augmented reality, and advanced sensory technologies, learners could explore immersive and interactive environments that transcend the physical realm. Students could manipulate virtual objects, conduct experiments in simulated laboratories, and collaborate with peers from different parts of the world in hyper-realistic settings. This policy decision would redefine the concept of the classroom and provide limitless possibilities for experiential learning.

Quantum Education:

Policymakers and academicians could harness the power of quantum computing and quantum information to revolutionize education. Quantum education would enable learners to engage in quantum simulations, algorithmic thinking, and quantum problem-solving. This policy decision would nurture skills such as quantum intuition, computational thinking, and data analysis at an unprecedented level. Students would gain a deep understanding of quantum phenomena and contribute to advancements in quantum technologies, shaping the future of science and technology.

Transdisciplinary Learning Pathways:

Academicians and policymakers could reimagine traditional subject-based curricula by designing transdisciplinary learning pathways. Instead of rigidly compartmentalized subjects, learners would engage in projects and investigations that cut across disciplines, integrating knowledge, skills, and perspectives from various fields. This policy decision would foster a holistic understanding of complex global challenges and promote innovative solutions. Students would develop the capacity to think critically and creatively, collaborating across disciplines to address real-world problems.

AI-Enabled Personal Learning Companions:

 Policymakers and academicians could introduce AI-enabled personal learning companions for every learner. These advanced AI systems would serve as personalized mentors, guiding and supporting learners throughout their educational journey. Equipped with sophisticated natural language processing, adaptive algorithms, and deep understanding of each student's strengths and weaknesses, these companions would provide tailored learning experiences, adaptive assessments, and instant feedback. This policy decision would empower learners to take ownership of their learning, continuously develop their 21st-century skills, and pursue personalized paths towards their goals.

Ethical and Emotional Intelligence Education:

Academicians and policymakers could prioritize the development of ethical and emotional intelligence as core components of education. This policy decision would emphasize the cultivation of empathy, compassion, and moral reasoning alongside academic knowledge and skills. Learners would engage in activities that promote ethical decision-making, conflict resolution, and social-emotional well-being. By nurturing these qualities, students would be equipped to address complex ethical dilemmas and contribute to a more compassionate and just society.

These wild and futuristic policy decisions may currently seem far-fetched, but they demonstrate the potential for transformative change in education and society. By pushing the boundaries of what is imaginable and challenging conventional thinking, academicians and policymakers can pave the way for a future where education and society exist in perfect harmony with the limitless possibilities of human potential. As we explore uncharted territories, let us remember that these wild ideas are meant to spark inspiration, ignite the fires of innovation, and challenge our preconceived notions of what education can be.

In this future vision, the rhythm of education pulses with excitement and anticipation. It is a symphony of minds interconnected, where the boundaries between physical and digital, individual and collective, fade into a harmonious blend of knowledge and imagination. Learners become creators, collaborators, and explorers, transcending the limitations of the physical world to embark on intellectual journeys that transcend time and space.

Imagine a world where learners, with their consciousness seamlessly transferred to digital realms, can embark on immersive learning experiences that defy the constraints of the physical body. Through mind-uploading and consciousness transfer, individuals can continue their pursuit of knowledge, creativity, and collaboration beyond the limitations of mortality. In these digital realms, learners interact with historical figures, engage in virtual debates with renowned scholars, and unlock the secrets of the universe, all while developing and honing their 21st-century skills.

Within these multi-dimensional learning spaces, holographic projections envelop learners, bringing concepts to life with breath-taking realism. Augmented reality overlays their physical surroundings with layers of information and virtual objects, allowing them to witness the microscopic intricacies of a cell or explore the vastness of the cosmos. Learners collaborate effortlessly with peers from different corners of the globe, transcending cultural, geographic, and linguistic barriers as they collectively unravel the mysteries of the universe.

As quantum computing and quantum information become commonplace, education transforms into an intricate dance with the fundamental fabric of reality. Learners engage in quantum simulations, where they harness the power of quantum algorithms and quantum mechanics to solve complex problems and explore new frontiers. With quantum education, students develop an innate understanding of the quantum realm, opening doors to unimaginable advancements in science, technology, and beyond.

Transdisciplinary learning pathways intertwine disparate fields of knowledge, as learners embark on quests that transcend disciplinary boundaries. By tackling real-world challenges, students explore the intricate connections between subjects, fostering a deep understanding of complex global issues and promoting innovative solutions. These pathways cultivate the 21st-century skills of critical thinking, creativity, and collaboration, empowering learners to become architects of a more sustainable, equitable, and interconnected world.

In this future landscape, AI-enabled personal learning companions serve as tireless mentors and guides, supporting learners on their educational journey. These advanced companions, infused with artificial intelligence, adapt to each student's unique learning style, preferences, and aspirations. Learners receive personalized learning experiences, tailored feedback, and adaptive assessments, enabling them to unlock their full potential and navigate the complexities of the ever-evolving world.

Ethical and emotional intelligence education takes centre stage, nurturing empathy, compassion, and moral reasoning in learners. By weaving these essential qualities into the educational fabric, students develop the capacity to navigate complex ethical dilemmas, foster inclusivity, and cultivate a deep sense of social responsibility. In this future society, individuals are not only equipped with knowledge and skills but also possess the wisdom and emotional intelligence to shape a better world.

As we venture into the realm of wild imagination, we must acknowledge that realizing these dreams requires boldness, collaboration, and an unwavering commitment to the transformative power of education. Academicians and policymakers must rise to the occasion, transcending the barriers of the present and embracing the limitless possibilities of the future. By fostering an ecosystem that nurtures innovation, encourages risk-taking, and embraces the unknown, we can cultivate a world where education becomes the catalyst for profound societal transformation.

In conclusion, let us continue to dream, envisioning a future where the rhythm of education resonates with the cadence of limitless possibility. These wild and futuristic policy decisions may seem distant and unattainable, but they serve as guideposts, illuminating the path towards an education ecosystem that embraces the full spectrum of human potential. With audacity and determination, let us march towards this harmonious future, where the symphony of learning reaches its crescendo, and the world dances to the beat of boundless imagination.

ABOUT THE AUTHOR

Dr. Vidhu P Nair is an esteemed educator, renowned for his remarkable contributions to the field of education. With a career spanning several decades, Vidhu has made a lasting impact through his expertise in curriculum development and educational leadership.

In 2018, Vidhu received the prestigious President's Medal for Teachers, a testament to his exceptional commitment to the art of teaching. Throughout his illustrious career, he has consistently demonstrated a deep passion for education and an unwavering dedication to his students.

Vidhu's influence extends beyond the classroom. He has played a pivotal role in shaping educational materials and policies as a contributor to textbooks and handbooks published by the National Council of Educational Research and Training (NCERT) in India. His meticulous work has not only enriched the learning experiences of countless students but also influenced educational reforms at the national level.

Moreover, Vidhu has actively contributed to various academic and assessment programs of the Department of Education and Department of Women and Child, Government of Kerala, and the Ministry of Human

Resource Development, India. His insights and recommendations have been instrumental in shaping the direction of educational policies and practices.

Vidhu's dedication to educational administration and planning is evident through his contributions to esteemed institutions such as the National Institute for Educational Administration and Planning (NIEPA), India, and the State Leadership Academy, Kerala. His expertise has helped nurture the next generation of educational leaders and equipped them with the necessary skills to drive meaningful change.

Beyond borders, Vidhu served in the State of Eritrea, Africa, as part of a World Bank-funded educational project for five years. His selfless commitment to improving educational opportunities for disadvantaged communities has left a lasting impact on countless lives.

A recognized researcher and thought leader, Vidhu has presented numerous papers at educational seminars and conferences in India and abroad. His innovative ideas in school administration and academic designs have earned him numerous awards and distinctions, highlighting his pioneering contributions in the field.

With humility and integrity, Vidhu continues to inspire educators, students, and policymakers alike. His forthcoming book is a testament to his vast knowledge and unwavering commitment, offering invaluable insights to those seeking to transform education for a brighter future.

www.ingramcontent.com/pod-product-compliance
Lightning Source LLC
LaVergne TN
LVHW052256210726
843527LV00041B/581